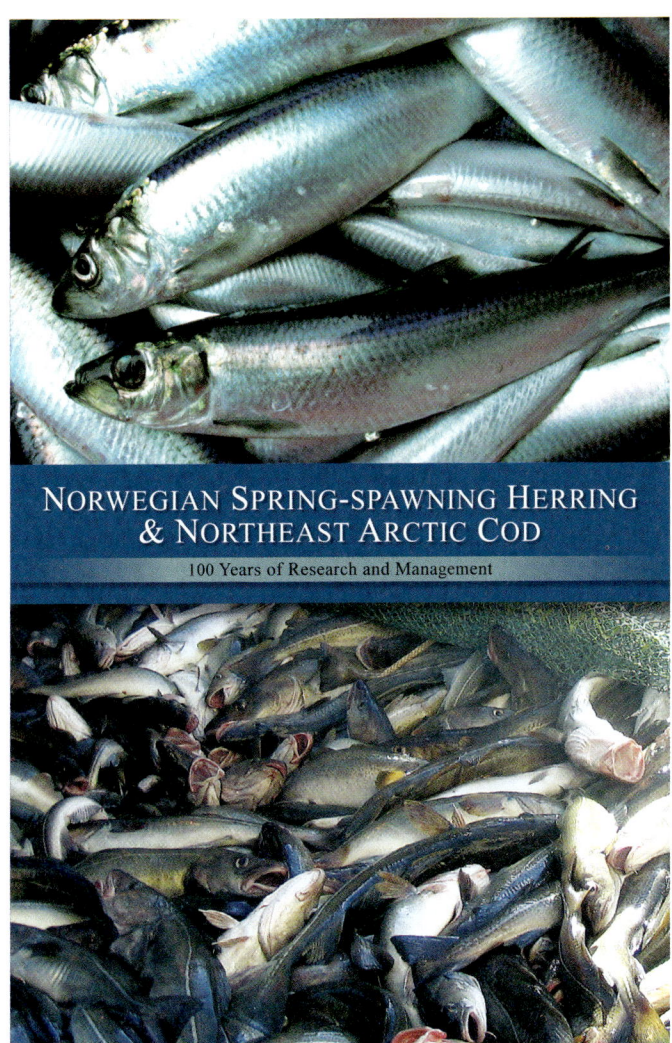

Norwegian Spring-spawning Herring & Northeast Arctic Cod

100 Years of Research and Management

Editor:
Odd Nakken

Illustrations:
Karen Gjertsen

Graphic design and technical editing:
Harald E. Tørresen

tapir academic press

© Tapir Academic Press, Trondheim 2008

ISBN: 978-82-519-2367-5

*This publication may not be reproduced, stored in
a retrieval system or transmitted in any form or by any
means; electronic, electrostatic, magnetic tape, mechanical,
photo-copying, recording or otherwise, without permission.*

Graphic design and technical editing:
Harald E. Tørresen, Institute of Marine Research

Printed by:
07 Gruppen AS

Photo cover:
Top: Jan de Lange
Bottom: Terje Jørgensen

Tapir Academic Press
NO–7005 TRONDHEIM
Tel.:   + 47 73 59 32 10
Fax: + 47 73 59 32 04
Email: forlag@tapir.no
www.tapirforlag.no

Contents

Preface 7

1. **Introduction** 9
   Odd Nakken

2. **From fisheries research to fisheries science, 1900–1940:**
   Tracing the footsteps of Johan Hjort in Bergen and the ICES 17
   Gunnar Sætersdal

3. **Technological developments in Norwegian fisheries** 33
   Steinar Olsen

4. **Norwegian spring-spawning herring:**
   history of fisheries, biology and stock assessment 41
   Olav Dragesund, Ole Johan Østvedt and Reidar Toresen

5. **Northeast Arctic cod:**
   fisheries, life history, stock fluctuations and management 83
   Arvid Hylen, Odd Nakken and Kjell Nedreaas

6. **The Barents Sea 0-group surveys;**
   a new concept of pre-recruitment studies 119
   Olav Dragesund, Arvid Hylen, Steinar Olsen and Odd Nakken

7. **Acoustics in fisheries science in Norway** 137
   Odd Nakken

8. **Fish behaviour, selectivity and fish-capture technology** 157
   Steinar Olsen

Preface

The production of this book was initiated by Gunnar Sætersdal in the mid-1990s. He suggested that we should write overviews covering the development in research and fisheries management throughout the 20th century of Norwegian spring-spawning herring and Northeast Arctic cod (also called Arcto-Norwegian cod and Arctic cod in the literature). Being the two most important commercial fish stocks for Norwegian and northeast Atlantic fisheries, these stocks were also among the first large marine stocks that were subjects to systematic research and monitoring. There exist almost unbroken time series of annual data on biological characteristics of the specimens in the catches from the beginning of the 20th century, for each of them; data that have been of major importance for the development of the advisory – and management system established during the most recent two-three decades. In addition, considerable amounts of knowledge has been gained by comparing the time series of stock characteristics with the 100 years long series of temperatures in the Barents Sea observed by the Polar Research Institute of Marine Fisheries and Oceanography (PINRO).

At the time Sætersdal proposed the writing of the book, both stocks were recovering from historically low levels. Stock sizes and yields were increasing largely due to effective fisheries regulations in the immediate past decade. As Director of the Institute of Marine Research (IMR) from 1970 to 1986, Sætersdal had strongly contributed to the research and management system generating these regulations. His idea was to have the book completed by the turn of the century when the Norwegian Directorate of Fisheries and the International Council for the Exploration of the Sea (ICES) celebrated their 100 years anniversaries. However, by his death in 1997 only two chapters (2 and 8) were written, and the work came to a stand still for many years.

I thank my colleagues (authors and co-authors), which have functioned as an editorial board. Special thanks to Karen Gjertsen for her work with the illustrations, Vibeke Kristiansen for typewriting, Ingunn Bakketeig and Elen Hals for quality assurance and corrections, Hugh Allen for language improvements. Harald E. Tørresen has done the technical editing and layout.

Bergen, June 2008
Odd Nakken

CHAPTER 1

Introduction

Odd Nakken

1.1 Fisheries and yields

Northeast Arctic cod and Norwegian spring-spawning herring have been the major target species in Norwegian fisheries throughout known history, and prior to 1960 catches of these two stocks made up 80 percent or more of annual Norwegian landings of fish. Until the mid-20th century, Norwegian fisheries were largely coastal and seasonal. They were based on the influxes of spawners of both species into near-shore waters in winter–spring as well as on the occurrence of feeding concentrations of immature fish later in the year; in April–June young cod feeding on spawning capelin along the coast of Finnmark and young herring, known as "fat herring", in summer and early autumn feeding on zooplankton in the fjords all along the coast. The fisheries for spawning cod have been off the counties from Møre and northwards, Lofoten being the most important fishing district, while the bulk of the spawning herring has been caught from Møre and southwards along the coast (Figure 1.1). In "Konungs Skuggsjaa" ("The King's Mirror"), a book written in about 1230, a father tells his son why fish are abundant in near-shore waters in early spring: "The fish come into coastal waters in late winter – early spring and spawn their roe so that the fry will have the long warm summer to gain strength in order to survive the next winter". The interest the authorities took in the cod and herring fisheries almost a millennium ago is shown by the facts that the king built lodges in Lofoten around 1120–1130 for the cod fishermen, and that the old Frostating law included rules on how to preserve herring.

A substantial increase in catches of both cod and herring took place during the 19th century; from about 50 000 to 200 000 tonnes of cod and from a few to about 60 000 tonnes of herring (Øiestad 1994). Catches continued to increase during the 20th century (Figure 1.2) until the stock of herring collapsed in the 1960s, while cod stocks suffered a major decline in the 1980s due to overfishing in the preceding decades (Hylen 1993). Improvements in fishing technology and rapidly growing international fishing fleets exploiting both cod and herring were the main factors behind the growth in catches up to the 1950s. The warming of the ocean water masses in the decades prior to World War II (1939–1945) also favoured production in both stocks in these years (Figure 1.2). Severe restrictions on fisheries – for herring there was a ban on fishing for several years – helped

Figure 1.1 *Counties, fishing districts and towns.*

both stocks to recover towards the end of the century. The main features of this development are described in Chapters 2–8, but first we shall bring readers up to date on some important background information.

1.2 Research; institutions, tasks, support and cooperation

Both cod and herring fisheries have always been subject to large fluctuations. Rollefsen (1966), who summarized the history of Norwegian fisheries research, wrote: "To the extent we are able to trace the Norwegian fisheries back through the ages we read of years when the fish approached the coast in great numbers, but more often of years when the herring and cod fisheries were failures". The large fluctuations in catches, and their effects on the economy of the country, stimulated the authorities to establish research programmes for both species in about 1860 (Rollefsen 1966; Solhaug and Sætersdal 1972), and these programmes led in 1900 to the establishment of the Norwegian research and management institution, the Direction of Fisheries, which in 1906 became the Directorate of Fisheries. Under the Directorate, fisheries research was organized in a special section which in 1947 was named the Institute of Marine Research (IMR), which since 1989 has been administratively separated from the Directorate of Fisheries.

During the first half of the 20th century, the main purpose of IMR's research was to provide answers to the question: Why do fish catches vary? And then, to advise on how, when and where to fish in order to improve yields. Along

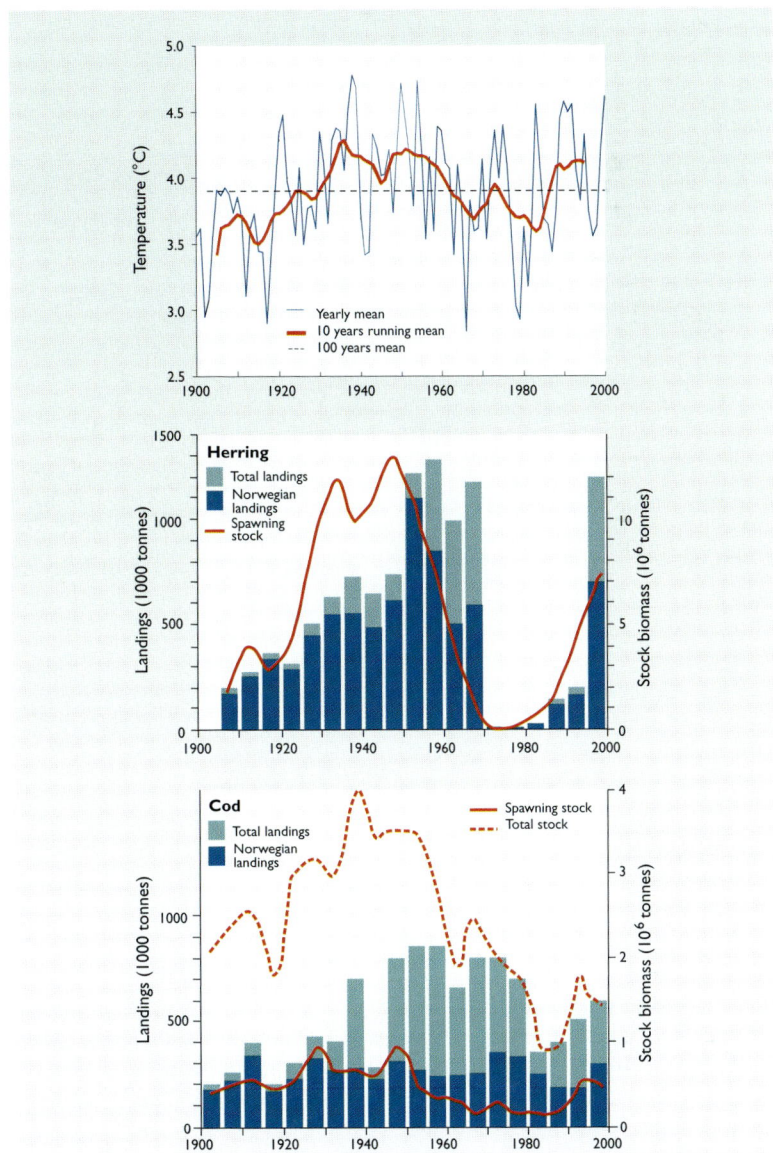

Figure 1.2
*The development in the 20th century of;
Upper: Temperature at depths of 0–200 m in the Barents Sea on the Kola Section.
Middle: Catch and spawning stock biomass of Norwegian spring-spawning herring.
Bottom: Catch, total stock and spawning stock biomass of Northeast Arctic cod.*

with the development of tools for assessing the impacts of fisheries on fish stocks and the acknowledgement that many stocks suffered from overfishing, including Northeast Arctic cod and Norwegian spring-spawning herring, fish stock assessments and advice regarding various regulation measures, including annual fishing quotas, became increasingly important. Throughout the century, cooperation within the International Council for the Exploration of the Sea (ICES) has been of great importance for the development of Norwegian fisheries science. Where the results achieved for the two stocks considered in this book are concerned, the cooperation with Icelandic and Russian scientists has been of immense value.

Until the end of World War I (1914–1918), Norwegian fisheries science was rather generously supported by the government. After the war, the situation changed (Rollefsen 1966). The nations that had been at war now fished their own fish in the North Sea where stocks had grown substantially during wartime, and the demand for Norwegian fish products sank. "Who wants marine research under such conditions?", quotes Rollefsen. However, in 1927, the Fishing Industry Research Fund was established, and "it must be said to the credit of those who had this Fund created that it has played a most important part in establishing Norwegian Fisheries research in the prominent position it now holds" as Rollefsen put it. Since World War II, fisheries science has again received generous support in Norway. The scientific and technical staff at IMR was strengthened substantially during the immediate post-war decade. In the 1950s, the institute brought two new and well-equipped oceangoing research vessels into use, in 1960, the staff of the institute moved into a large new building, and throughout the 1960s and 1970s, new research groups were established and new research vessels built in order to meet the growing demands for knowledge of marine resources and their environment. At the end of the century, IMR's research profile included most aspects of importance for the production from commercial fish stocks as well as from Norwegian marine aquaculture.

In the early 1970s, the capacity for doing fisheries science in Norway was also considerably strengthened through the establishment of several new research institutions: The Institute of Fisheries Technology Research in Tromsø with its Fish-Capture Division situated in Bergen, and the Norwegian College of Fisheries Science in Tromsø, with its Institute of Fisheries Biology also situated in Bergen (since 1980 as part of the University of Bergen). Both of these institutions were located in Bergen in order to facilitate cooperation with the Institute of Marine Research; a cooperation that has been most fruitful where Norwegian fisheries science and education are concerned. Funding was substantially increased through the establishment of the Fishery Research Council in 1972.

1.3 Fishing limits and ownership of stocks

The importance placed on protection of her coastal fisheries, particularly the cod fisheries in the north (Finnmark, Lofoten), led Norway to claim a four nautical mile territorial limit (rather than the conventional three nautical miles) as early as the 19th century. Baseline positions for the fishing limits were selected so that important traditional fishing grounds became territorial waters in order to prevent foreign fishing vessels from operating there. When the international trawl fishery developed in the Barents Sea area early in the 20th century, Great Britain,

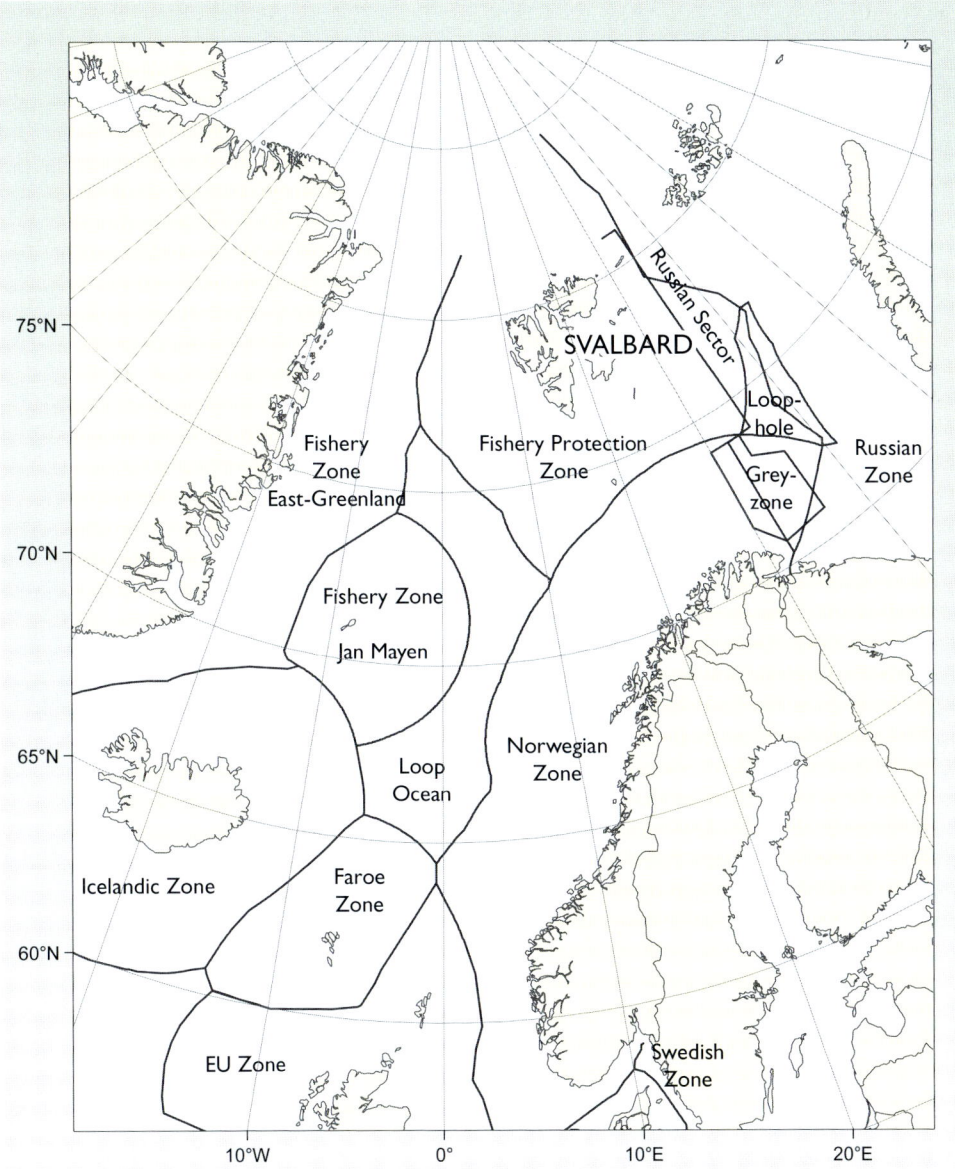

Figure 1.3 *Fishing limits established in 1977.*

then the leading trawling nation, disputed the Norwegian territorial claim. After long negotiations, the dispute was settled in 1951 by the International Court of Justice in The Hague, which approved Norway's claims.

In 1961, a 12 nautical mile fishing limit was established. Even so, Norwegian authorities felt that further protective measures of coastal fisheries were needed, and in 1973, several seasonal "no trawling zones" were established outside the 12 mile limit of northern Norway. This was an interim measure pending the establishment of a 50 nautical mile exclusive zone which by then had become a public demand in the northern counties. However, Norway claimed a 200 nautical miles economic zone effective from January 1977, in accordance with the extended coastal state jurisdiction under the Law of the Sea regime (Figure 1.3). In June that same year, a 200 nautical mile "fishery protection zone" was established around the Svalbard Islands in order to control fishing activity on the important nursery and feeding grounds for cod in that area. The Svalbard protection zone differs from an exclusive fishing zone in its requirement to treat all signatories of the Svalbard Treaty equitably. Norwegian authority in this zone is challenged by other states. Russia (USSR) did not recognize the protection zone, but practical arrangements between the two countries for orderly fishing in the zone have proved to be possible.

The boundary between exclusively Norwegian and Russian waters in the Barents Sea is not agreed. Norway claims an equidistant line from the coasts while Russia claims an "Arctic sector" with a western boundary along the 31°30'E meridian. As a practical solution a "grey zone" was established in 1978, pending a solution of the boundary issue. Within the grey zone each party has jurisdiction over its own fleet, and, within a joint framework, may licence fishing by third parties.

An area in the central Barents Sea, known as the Loophole, is outside the 200 nautical mile zones of both Norway and Russia. At times, particularly in warm periods, fishable concentrations of cod are found here. In the mid 1990s, Icelandic vessels caught substantial quantities in the area and thus forced Norway and Russia to accept that Iceland should have a share of the annual total allowable catch (TAC) for Northeast Arctic cod. Part of the Norwegian Sea is also outside national waters, and at times, catches of herring in excess of the agreed TAC have been taken there.

Many fish stocks are transboundary, including Northeast Arctic cod and Norwegian spring-spawning herring, and the new coastal state jurisdiction made it necessary to establish shared ownership of the stocks and a system for sharing the total TAC. In 1975, Norway and Russia established a forum for cooperation and negotiations regarding fisheries management in the Barents Sea area, the Mixed Norwegian-Russian Fishery Commission. For Northeast Arctic cod, agreement was reached on allocating fifty percent ownership to each of the two countries. For herring, which are distributed throughout large parts of the Norwegian and Barents Seas, Norway, the EU, the Faroes, Iceland and Russia annually agree on the TAC and its allocation to individual countries, but have not yet reached agreement on stock ownership on a permanent basis.

1.4 Overview

In this book we describe the development of the fisheries and stocks of Norwegian spring-spawning herring and Northeast Arctic cod during the 20th century, including the impact of environmental factors and fisheries on this development, and the measures that have been recommended and introduced in order to conserve stocks and maintain yields at reasonably high levels. Research in fisheries acoustics and fish-capture technology has contributed significantly to the results obtained in the second half of the century, and we have therefore included descriptions of developments in these two fields.

Chapter 2 deals with the efforts and influences on fisheries science of Norwegian scientists early in the century, in three main fields:

- Their utilization of research surveys to describe the life history of stocks.
- Their discovery of the large variations in year-class strength, and their struggle to establish an international (ICES) database of stock age compositions; i.e. catch at age figures for each fish stock, data which many years later became the "backbone" of analytical stock assessment.
- Their contributions to the understanding of the effects of fishing and the formulation of a theory of fishing.

Chapter 3 summarizes the technological developments in the cod and herring fisheries and in gear and vessel technology; developments that made fishing so efficient that due to absent and/or ineffective management, the stock of herring was in a state of collapse for 20–25 years, while the cod stock was at an extremely low level for about a decade (Figure 1.2).

Chapters 4 and 5 present and evaluate current information on trends and fluctuations in fisheries, biological characteristics and stock biomass for the two stocks. The observation and sampling systems that have been developed and used to monitoring stocks are described, and the management measures which have been recommended and introduced are outlined.

Chapter 6 is a description of the methodology employed and results obtained in the Barents Sea 0-group survey. This survey set a standard for the conduct of acoustic surveys for measuring fish abundance. It has been carried out every year since 1965 and has provided an early indicator of the strength of cod and herring year-classes as well as of many other fish species; data that have been used in numerous studies of fish recruitment.

The ability to find fish and study their distribution and behaviour and to measure their abundance was greatly improved when acoustic instruments came into use in the mid-1930s. During the second half of the century, Norwegian scientists and engineers were at the forefront of improvements in acoustic instrumentation and methodologies for fisheries science; developments that are dealt with in Chapter 7, as well as in Chapters 3, 6 and 8. The echo integration technique, which has become a standard method for measuring fish density and abundance, was invented by Norwegian scientists and used by them on 0-group fish in the Barents Sea in the early 1960s.

Chapter 8 deals with studies of fish-capture technology and fish behaviour related to capture and abundance estimation. In Norway, this type of research enjoyed growing support in the 1970s, since when continuous efforts have been made to enhance fish capture technology in the commercial fisheries, to diminish by-catch of undersized fish and non-target species, and to reduce bias and improve the accuracy of surveys and abundance estimates made from research vessels.

REFERENCES

Hylen, A. 1993. Impact on marine fish populations. In: Sundnes, G. (ed.): Human Impact on Self-recruiting Populations. The Kongsvoll Symposium 1993. The Royal Norwegian Society of Sciences and Letters Foundation. Tapir Publishers, Trondheim, Norway.

Rollefsen, G. 1966. Norwegian Fisheries Research. Fiskeridirektoratets Skrifter, Serie Havundersøkelser, 14(1): 1–36.

Solhaug, T., Sætersdal, G. 1972. The Development of Fishery Research in Norway in the Nineteenth and Twentieth Centuries in the Light of the History of the Fisheries. Royal Society of Edinburgh, Proceedings B: Biological Sciences, 73.

Øiestad, V. 1994. Historic changes in cod stocks and fisheries: Northeast Arctic cod. ICES Marine Science Symposia, 198: 17–30.

CHAPTER 2

From fisheries research to fisheries science, 1900–1940: Tracing the footsteps of Johan Hjort in Bergen and the ICES[1]

Gunnar Sætersdal

2.1 Introduction

A few years after the Norwegian Parliament had created a new government organization; the Direction of Fisheries in Bergen in 1900, this institution had developed into a centre for comprehensive programmes of investigations of the basis of fisheries, fish stocks and their physical environment. Its studies were directed mainly at problems in the great Norwegian fisheries for Arctic cod and spring-spawning herring, but were soon extended to other stocks. The results often proved to be of great general interest.

The period 1900 to about 1914 has been termed the golden age of Norwegian fishery investigations (Rollefsen 1962). To an unusual degree the advances made in this golden age can be ascribed to the innovative ideas and the dynamic personality of one scientist, Johan Hjort (Schwach 2002). European fishery research at this time was undergoing an important process of internationalization, and Hjort became a central figure in that process. His steps thus lead us through the main events in the early development of both Norwegian and European fishery research, as well as those of the creation and early history of the International Council for the Exploration of the Sea (ICES).

One of the main practical objectives of ICES; to function as an advisory body for the regulation of fisheries, would prove to be a long-term goal that would not be reached until the 1930s. This slow progress no doubt reflects that of fishery science, with the first true insights into the reaction of fish stocks to exploitation dating from the mid-30s. But much happened in the meantime and we shall try to describe this eventful period for fisheries research step by step.

2.2 Early objectives and programmes of the Bergen institution
Research surveys
In his proposal to the Home Office in 1899 for the building of a research steamer for Norwegian fishery research, Hjort's main argument was the need to extend fishery research beyond coastal waters (Hjort 1899). There was hardly any information on the location of the large masses of cod and herring after they

---

1) A somewhat shorter version of this article is published in ICES Marine Science Symposia, 215: 515-522. 2002.

left the coast following spawning. There was a clear need for an expansion of the Norwegian fisheries which were still mainly coastal, and the potential was indicated by the growth in the previous decade of the trawler fleets of Great Britain and Germany which had left the crowed North Sea grounds for better conditions around Iceland, Spitsbergen and even in Norwegian coastal waters. In broad terms, the objectives were to describe all aspects of the natural history of the cod and the herring, study their environment, and explore conditions for expansion of the fisheries to offshore and distant-water grounds (Hjort 1909).

The choice of the type of vessel was unconventional and undoubtedly of great significance for the success of the programmes. Hjort recommended that she should be built "exactly like one of the modern fishing steamers ..., and make use of all the fishing appliances of the present day in the service of science" (Hjort 1909).

This simple principle of using a vessel which, in experiments, could simulate commercial fishing would greatly enhance the value of the observations obtained and is today recognized in the design of most modern fishery research vessels. However, at the time it was a new idea. Discussing the work done with the MS "Michael Sars" in the period 1900–1908, Hjort (1909) said: "So far as I know she is the first to be built like one of these fishing steamers which has experimented with all modern methods of ocean research. Now there are several vessels like her, and it seems to be recognized that they denote a great step forward."

Commissioned in July 1900 after only a year spent on planning and building, at 226 GRT the "Michael Sars" was not a large vessel by modern standards, about half the size of the present "Michael Sars", the smallest of the current fleet of the Institute of Marine Research. She was deployed in surveys which covered the Skagerrak, the northern North Sea, the Norwegian Sea and the southwest Barents Sea. The work was intense, and cruises were made in all seasons of the year in spite of unfavourable weather conditions in the winter, which were even worse than had been expected.

The inauguration cruise to the Norwegian Sea, Iceland, Jan Mayen and Bear Island focused on the pelagic system over the deep-sea areas, at depths in excess of 400 m (Hjort 1901). A remarkable finding of this first survey was the wide distribution of drifting juvenile fish in the warm Atlantic water, and Hjort discussed the possibility of obtaining estimates of their abundance. Some 60 years would pass before this approach was made use of in the important 0-group surveys started by Dragesund and collaborators off northern Norway (see Chapter 6).

The "Michael Sars" surveys were in support of a general research programme with the following elements (Hjort 1909):
- General ocean research, consisting of hydrography and plankton investigations.
- Investigations into the natural history of fishes (the chief scientific task). For these objectives the surveys were directed towards the various stages of the fish: eggs and larvae to chart the spawning grounds; pelagic and bottom stages of the post-larvae and juveniles, and adults. The systematic collection and analysis of these data would help to describe the distribution, migrations and movements of the fish and thereby help identify the stocks as well as provide information on size and growth.

The fishery research vessel "Michael Sars".

- Fishery experiments to test the potential of offshore grounds for an expansion of the Norwegian fisheries.

The tools and gear used were hydrographic instruments, Hensen nets for vertical plankton hauls, larger plankton nets (up to 7 m in diameter) for towed hauls at various depths for post-larval stages, a 120-foot bottom trawl, longlines and drift nets.

The novelty of these survey programmes of the Bergen institution lay in their broad and ambitious approach. They were directed at all stages of the fishes; eggs and larvae, pelagic and demersal juveniles and adults. Catches of adult fish were comparable to commercial catches both in terms of catch rates and size compositions. Relevant commercial data could thus be included in the analyses. The surveys also covered wide ocean areas which encompassed the distributions of nearly all stages of the fish stocks on the Norwegian shelf.

The results were reported with little delay by Hjort and his co-authors. The first survey is dealt with in Hjort (1901), the first four years of work in Hjort (1905) and the period 1900–1908 in Hjort (1909). The final report of ICES Committee A (ICES 1909a), described the results of the Bergen investigations together with those of the other cooperating countries. Finally the survey results formed important parts of the database for the now classical "Fluctuations in the Great Fisheries of Northern Europe" (Hjort 1914).

This period of intensive studies by the Bergen institution from 1900 until the outbreak of World War I produced great advances in our knowledge of the seas around Norway and their resources. The main effort was directed towards Arctic cod and Norwegian spring-spawning herring, and the progress made in revealing the life histories of these two stocks is reviewed in Chapters 4 and 5.

In 1914, the "Michael Sars" was requisitioned to safeguard Norwegian neutrality. She was never again used for fisheries research, and the Bergen institution lost its most important tool for studying these two highly migratory stocks. Thus, Norway's studies of cod and herring did not continue the remarkable progress of the first "golden age". Not until the 1950s would the Institute of Marine Research be able to resume the once successful investigations by means of wide-ranging surveys.

Stock age compositions

In addition to the descriptions of the life histories of fish, Hjort and his collaborators made a fundamental contribution to fishery science, through their utilization of representative age compositions of stocks as a tool to study stock fluctuations, stock identity and vital stock processes.

Examples of the use of structures in scales, otoliths and bones for age determination of both fresh water and sea fishes were well known by the turn of the century. The first use of this method by the Bergen group was apparently made by Hjort's assistant Hjalmar Broch, who in 1904–1906 made a special study of European herring races, primarily employing Heincke's method (Broch 1908). He compared scales of herring from a number of different localities and found distinct structures which he assumed were the result of seasonal variations in growth. Dahl (1907) and Lea (1910) continued the studies of herring scales, and Lea, over a period of 15–20 years, became deeply involved in investigating them, especially their use for estimating age and growth and as a certificate of origin (Ruud 1971). Age determination by scales was soon also used for other species, especially cod and haddock.

The first and most remarkable results of these investigations, which included a very large number of samples from the North Sea and the Skagerrak up to the Barents Sea, were that the age groups for some species, e.g. haddock and herring, showed similar characteristic patterns in samples over very large areas (Hjort 1907). It was thus possible to describe the fish stocks of certain areas by age "censuses". The dimensions of the year-class fluctuations showed them to be significant natural phenomena, unrelated to the fisheries. The prospects of being able to put whole fish stocks into age-class systems reminded Hjort of human population studies, and he suggested, in a lecture at the 1907 ICES meeting, that fishery biologists might find it useful to think in such terms (Hjort 1908).

Some of the data for the stock age composition studies were from the programme of ICES Committee A. Although this cooperative exercise ceased in 1908, the work of the Bergen group continued with material and data from Norwegian herring and Arctic cod. These observations gave Hjort and his group sets of age compositions for both of these stocks which, in a most convincing way, demonstrated the validity of the method and its practical application. The year-class of 1904 was unusually abundant in both stocks, an observation which was repeated year after year (Lea 1929). This series not only confirmed that the age determination must have been correct in a very high proportion of the cases, but also that the system of sampling had given meaningful and repeatable results. Samples of age compositions of juvenile fat herring and adult winter herring also demonstrated meaningful passages of strong and weak year-classes. A further type of information regarding the herring stock was that where it had spent juvenile life stages could be identified from the evidence of slow or rapid growth laid down in the patterns of rings of the scales, making them certificates of origin.

In his "Fluctuations in the Great Fisheries of Northern Europe", Hjort (1914) delivered a message, saying: "By large-scale stock age compositions one can identify stocks, classify them by age, measure their growth and make predictions of changes in their fishable biomass". He also suggested that these types of data could be used in some form of vital statistics similar to those used in human population studies. This was a clear and powerful communication, but its reception was disappointing.

Sinclair and Solemdal (1987) found that Hjort's introduction of the age compositions of stocks was significant in the development of population thinking among European fishery scientists, and that most of them had adopted that concept by 1920. However, this appears to be a rather limited utilization of the tools offered by Hjort's findings. As we shall see below, as we follow the development of fishery science within the ICES community, a considerable delay seems to have occurred between Hjort's presentations in 1908–1914 of what we may term his breakthrough in fishery biology and its acceptance and actual employment by other European fishery scientists, especially the British. In 1929 and 1930, ICES called two special meetings on year-class fluctuations (ICES 1930a, 1930b) where the subject was approached almost as if the concept was a new one. There was thus a delay of 15–20 years in the reaction of the community of fishery scientists, at least as represented by ICES, to Hjort's spectacular findings.

2.3 The creation and early history of ICES

Since Hjort was a keen supporter of ICES and an active operator in its affairs from the start, we can still follow his footsteps for a broad description of its activities, in particular of the Council's role in matters relating to fisheries research and fisheries.

At the first preparatory conference in Stockholm in 1900, detailed recommendations concerning international hydrographical studies were presented, while plans for biological research were only referred to in general terms. This may have reflected the priorities at the time of some of the influential founding fathers such as Otto Petterson and Fridtjof Nansen. The records show, however, that the British delegation was instructed to assign most importance to practical fisheries research, as was demonstrated by the wording of the introductory paragraph to the list of resolutions.

This paragraph was repeated in the preamble to the recommendations for cooperative investigations which appear in the report of the second preparatory meeting held in Kristiania in 1901 (ICES 1901), and it must indeed be said to give high priority to fisheries research:

"Considering that a rational exploitation of the seas should rest as far as possible on scientific enquiry, and considering that international cooperation is the best way of arriving at satisfactory results in this direction, especially if in the execution of the investigations it be left constantly in view that the primary objective is to promote and improve fisheries through international agreements ...".

For the second preparatory conference in Kristiania, Hjort contributed a conference document (ICES 1901) that described two main problem areas: 1) the possible overfishing of local stocks, and 2) the causes of the periodic occurrence of migrating fish. These were to be the research objectives to be dealt with by the ICES Committees A and B. His proposals and plans for the international investigations were a description of the survey programme already started by the Bergen institution, which specified special studies of the various stages of the fish: egg, larvae and post-larvae by means of plankton surveys, special young fish surveys with research vessels, while for the adult stages a combination of commercial statistics and experimental fishing would have to be used.

A further demonstration of Hjort's attempts to have his new approach to fisheries research adopted in the coming international collaborative studies is demonstrated by recommendation D of the Protocol of the Kristiania Conference: "The Conference considers it absolutely indispensable that each of the countries concerned should provide a steamer specially constructed for scientific fishery researches". Hjort's enthusiasm for survey methods in fishery research no doubt stemmed from the promising results of the start of the "Michael Sars" operations. This methodical approach is interesting in a historical perspective since programmes based on surveys came to be the most important element of IMR's resources research from 1950 onwards. By that time, however, the prominent use of this expensive research method was hardly a choice by tradition, but rather related to the special characteristics of Norway's cod and herring resources, stocks which migrated over large expanses of the sea.

At the inaugural meeting of ICES in Copenhagen in 1902, Hjort announced that according to the proceedings of the Norwegian Storting, the grant-in-aid for the international study of the sea was made for the express purpose of obtaining practical results, and that he believed this was the case also for the contributions of some other countries. In view of these limitations it appeared to be necessary to concentrate attention on a few important problems of practical interest. These were later adopted in nearly identical form as the terms of reference for Committees A and B, and are thus worth quoting:

1. The migrations of cod and herring and the influence of these migrations on the fisheries, especially in the northern part of the North Sea, and also the biology of these and other allied species.
2. The question of overfishing, particularly in the southern North Sea, and in connection with this, the special study of flatfish.

The committees were to conduct their investigations over a three-year period (later extended to five years) and then report back to the Council.

In the process of the creation of the International Council for the Exploration of the Sea several nations had, as noted above, stated their views that practical fisheries research should be the Council's first priority. The preamble must be interpreted as describing a regional body for cooperation in fishery research and for providing advice on the regulation of international fisheries. With three-mile territorial limits also representing the limits of coastal state jurisdiction over fisheries, in practice all main sea fisheries were international. The reports of the Committees were doubtless keenly anticipated, but we can now see that 30 to 50 years, depending on the standards set for advice, would pass before ICES would start to function in an advisory capacity with respect to the regulation of

Johan Hjort on board "Michael Sars" in 1902.

fisheries. This was obviously caused by the slow progress of the understanding of the nature of fish populations and their reaction to exploitation in the period.

The most important achievement of ICES in these early years was the organization of the cooperative international investigations through the Committees, later called Commissions.

Hjort was elected chairman of Committee A which had representatives from the UK, Sweden, Denmark, Russia and the Netherlands. At its first meeting in 1902 (ICES 1903), he proposed that work in the North Sea should be undertaken on the lines of investigations made on the Norwegian coastal banks; viz. fishing surveys on the banks in summer and winter, combined in winter with plankton surveys for eggs and/or larvae. In this way the migrations and movements of the adult fish between winter and summer would be shown, as well as the locations of the spawning grounds. Special survey efforts would have to be aimed at the young stages. In a progress report in 1905 (ICES 1905), Hjort refers to particularly good material on the young stages of Gadoids in the North Sea obtained from research vessel surveys, with German and Danish vessels operating in the southern North Sea and "Michael Sars" in the northern part.

2.4 The introduction of stock age compositions in ICES

There was some uncertainty regarding how long the Committees were to function. Originally set at three years, they were extended to five years, but in 1907 proposals were made for further extensions. By this time the Bergen group had produced the first stock-age compositions, and Hjort proposed to the members of his Committee that even if there were no binding tasks in the coming year, the main objective of future investigations should be annual age censuses of the principal food fish in particular regions (ICES 1907). Similarly, for the last meeting of the Committee, which took place in 1908, mainly for discussion of the final report, Hjort had circulated in advance a document containing a proposal for a programme of future work (ICES 1909b). For Gadoids he recommended the continuation of the present programme, including egg surveys, but if funds were scarce, that attention should concentrate on age sampling of fish on the largest possible scale. For herring too he recommended systematic collection of material for age and growth studies, but also pointed to the possibility of studying the distribution of different populations of herring by using similar methods.

Committee A accepted the Bergen Group's combined survey programme as a basis for its work in the North Sea, and as soon as the stock-age composition breakthrough was made in 1907, Hjort did his best to have also that approach adopted in the North Sea. For financial reasons, however, the cooperative investigations were not continued. In the introduction to his 1914 opus, Hjort stated that at the time he had very much regretted the discontinuation of the work of Committee A in 1908 and still did so. This represented a lost opportunity to test his approach in other areas and for other stocks.

The 1909 Committee A report was a very comprehensive document that contained a wealth of new information on the Gadoids from the North Sea and other parts of the North Atlantic (ICES 1909a). In the summary, Hjort discusses "Some practical fisheries questions in the light of the results obtained". There was evidence of distinct year-class fluctuations in the North Sea haddock, and Hjort again made a strong plea for future representative statistics that should include length measurements and age determinations in the data collected. It had been demonstrated that commercial size-classes (used for analysis by statisticians) contained several age groups, and it was thus impossible to demonstrate any regularity whatsoever in the changes or fluctuations of these classes, while the study of age groups could lead to factual insights into the question of the relation of fisheries to the stock of fish. This was no more than the truth, but it must have sounded like very plain language to the ears of D'Arcy Thompson, who used commercial data in his contribution to the report.

The part of the report of interest for possible fishery regulations deals with by-catches of undersized fish. Hjort found that "It is permissible to think that in taking the young fish before they have reached any mentionable value, the fishermen are treating the annual natural increase in an uneconomical manner". The discards in the cod and haddock fisheries were substantial, and it was concluded that large quantities of cod and haddock were taken at sizes which had only very small market value. The answer to the question of whether these Gadoid stocks were overfished would have to await further investigations.

Progress reports on the herring investigations begun in 1909 were soon available (Hjort 1910; Lea 1910) and were submitted for discussion at the Council meeting in 1910. Lea's paper was a serious treatise on the scale method, but D'Arcy Thompson criticized the age and growth investigations on the grounds that the number of rings were subject to individual variation. When Hjort reported on further progress in the herring investigations at the 1912 Council meeting

(Hjort and Lea 1911), Thompson again regarded age determination based on the number of scale rings as hypothetical. The full confrontation came at the 1913 Council meeting (ICES 1913). In response to lectures to the Council by Hjort and Lea, Thompson stated that he considered the method of age determination by scales to be //in proves//merely provisional, and could therefore not agree with the views of Hjort's lecture. Hjort then declared that no programme for herring research could be drawn up before this question was settled.

This dispute over the herring scales lasted more than 10 years. According to Wendt (1972), it was probably the first serious rift in the Council since its inception. With a perspective today of more than half a century's general use of the methods disputed by Thompson, it is difficult to understand the position he maintained for so long, especially since there is no documentation in his name to substantiate it. One cannot avoid the conclusion that Thompson must have needed a good portion of academic arrogance to be able to persist in his error for such a long time.

The dispute is now only of historical interest as an incident in the development of fisheries research. Lea claimed the dispute set back herring research several years (Wendt 1972). However, it is likely that the effects were wider and longer lasting. As noted above, there was a remarkable lack of response to Hjort's repeated "messages" over the years 1907–1914 of the importance of stock age compositions. Although Hjort seems to have carried the rest of the ICES representatives in the herring scale dispute, Thompson had been British delegate to ICES since 1902 and had great prestige and influence in Britain where several scientists supported his stance. And in the period between the two world wars, the Lowestoft and Aberdeen laboratories were about to take over as leading European institutions in fisheries research after the Bergen group.

The herring scale dispute was settled by a joint practical exercise in Kristiania in 1923, after which British herring scientists adopted the scale method. D'Arcy Thompson's final year as British delegate was 1925. A new generation of British fishery scientists took an interest in stock age compositions and population theories in the advancement of a fishery science.

2.5 Towards understanding the effects of fishing

Recalling that the purpose of this review is to trace the progress of fishery science for a better understanding of the history of our studies of the stocks of Arctic cod and Norwegian herring and their international role, the account of this period has largely been based on and restricted to ICES events and documentation.

As mentioned above, the 1929 ICES special meeting on "Fluctuations in the abundance of the various year-classes of food fishes" (ICES 1930a) must be seen as the first reaction by European fishery scientists to Hjort's messages of 20 years earlier. In his introductory address to the meeting, Hjort welcomed the great interest shown by the submission of 19 papers and noted the great advance made from the position 30 years earlier when scientists worked at species level. He foresaw that systematic international observations of year-class fluctuations would shed new light on many questions concerning fish production. One innovative contribution which demonstrated one of the most important future uses of age determination was Lea's "Mortality in the tribe of Norwegian herring". The emphatic statement from one of Hjort's old disciples, Oscar Sund, that age determination in the Arctic cod is "utterly unreliable except for the earlier age groups" created some confusion. We may wonder why, in the course of some 25 years of studying this stock, Sund stubbornly kept to scales and did not try the otoliths that had proved so useful in other fish. In an analysis of fluctuations in the trawl fishery for North Sea cod, D'Arcy Thompson seemed to be unsure whether they were caused by year-class fluctuations, while Harold Thompson was certain that the very large variations in the abundance of North Sea haddock were caused by fluctuations in the original brood strengths.

A follow-up meeting on the same subject was called in 1930 to discuss the reports prepared by special council nominees to the 1929 meeting, indicating the main results brought out by papers read at the 1929 meeting (ICES 1930b). In a preface, E.S. Russell pointed out the main problems to be addressed: the validity of the method, the organization of regular cooperative studies of the fluctuations, and investigations of their causes. In his notes for the meeting, Michael Graham, the reporter for cod, maintained doubts about the age determination of older fish, referring to Sund and others. He saw the importance of early observations of brood strength for forecasting stock fluctuations, and recommended continued assessment of smallest marketable size groups and observations of pelagic fry – an early recommendation for 0-group surveys. As the reporter for herring, Lea stressed that further racial studies were needed as a basis for reliable stock fluctuation estimates. He called for standardization of methods and foresaw particular problems in obtaining representative sampling from seasonal fisheries. On the causes of fluctuations he referred to Hjort's hypothesis.

The two meetings once again drew attention to the phenomenon of year-class fluctuations, and drew out some new data, but apart from Lea's mortality estimates, did not reveal any advances in methods or thinking. Apparently little had happened in this field for about 20 years, although Norway continued studies in support of the original work of the Bergen group, with several herring papers by Lea in the 1920s. Particularly elegant was the paper "The herring scale as a certificate of origin" (Lea 1929) which demonstrated his idea of the scale as the identity card of the herring. There is no doubt that Lea emerged as the most distinguished scientist of the Bergen group after Hjort.

A few years after the ICES special meetings on year-class fluctuations, another important contribution to the subject came from Norway; Rollefsen's paper "The otoliths of the cod", in which he showed that the problem of estimating the age composition of the stock of skrei could be solved and that estimates could be made of the spawning class composition of the stock (Rollefsen 1933). This was a great methodical advance that must have comforted Hjort and may even have removed Michael Graham's lingering doubts.

In 1932, ICES called a special meeting on "The Effect upon the Stock of Fish of Capture of Undersized Fish" (ICES 1932) The nine papers presented did not contain any clear assessments of the effects, but after a discussion it was concluded that the existing rate of capture of younger age groups of haddock was extremely undesirable and that considerable savings could be made by adopting larger trawl meshes. For plaice it was considered that an unrestricted fishery on the nursery grounds was a danger to the maintenance of a marketable stock and was likely to be followed by a decline in yield.

Without actual assessments, the position of ICES was not strong. Hjort made a special contribution to this meeting which seems to have described how most fishery scientists must have felt at the time: the small fish question was as old as the knowledge that the catches of the North Sea trawlers largely consisted of undersized fish. This problem had been intensively studied, and proposals for "avoiding the enormous waste"; size limits, closed areas and savings trawls had been submitted. However, until then, these had aroused only academic interest, partly owing to practical difficulties, and partly because of the resistance shown by the fishing industry and governments to accepting any limitations on their unrestricted freedom of action.

There had also been opposition on theoretical grounds that the growth of North Sea fish was inversely related to the number of fish. This "thinning theory" was closely connected with old biological theory and experience dating back to Malthus and Justus von Liebig, who showed that populations are dependent on "means of subsistence". Modern marine biological research had also shown a tendency to assume that a population always utilizes existing means of subsistence and was, therefore, always proportional to these.

Hjort showed examples of Norwegian herring with 30-fold changes in year-class abundance but with no change in growth rate. Iceland haddock and plaice in Danish waters also showed about the same growth rate for year-classes of different sizes. He concluded that he was unable to abandon the leading idea which had guided ICES from the beginning; that the enormous destruction of young fish in the trawl fisheries should be prevented. This was the state of the art in the early 1930s. No convincing assessments of the effects on stocks of the capture of undersized fish had been made, nor had it been clearly demonstrated that these captures endangered the stocks or caused declines in yields. But there was an obvious risk that this might be the case, and the meeting followed what today would be called the precautionary principle of management advice. This was sensible, but it was not the basis of a strong case which could persuade governments and industry.

The 1934 special ICES meeting on "Size limits for Fish and Regulation of the Meshes of Fishing Nets" (ICES 1934), addressed the practical questions of how to limit the capture of undersized fish, the need for which had by now been agreed. The by-catch of undersized fish had been perceived as the main "overfishing" problem since the start of Committee B. This was an unfortunate limitation of the concept of overfishing, inherited from more than 30 years of discussing the problems of the plaice fisheries. The two special ICES meetings that dealt with this problem led to the 1937 London Conference, which focused the attention of the fishery regulatory authorities on mesh size and landing size for fish, attention which was maintained well into the 1950s, although Britain attempted to start discussions of limitations on catch effort immediately after World War II.

2.6 Formulation of a theory of fishing

Parallel to these efforts on fishery regulation, which as noted above seem to have had their basis in the use of the precautionary principle, several important contributions were made towards the formulation of a theory of fishing. E.S. Russell (1931) formulated the parameters of an arithmetic function which described yield as a function of recruitment, growth and natural and fishing mortalities. This identified the important processes, but the practical usefulness of the function was limited by the partial interdependence of the parameters. In their paper "The optimum catch", Hjort et al. (1933) referred to the effects of the War on the North Sea stock of plaice, and postulated that it is perfectly possible for an intelligent community to create an industry based on an optimum catch. That Europe had not done so was due to the lack of restrictions on the fishing industry. In the theoretical part of the paper, an analogy was drawn between a stock in the sea and a population of yeast cells in a closed vessel. The rate of increase is initially small, attains a maximum, and then decreases to zero. On this basis it is possible to estimate theoretical curves representing the dynamics of stock and regeneration and thus to deduce whether the stock will support the catch. These are sigmoid curves, and the point of inflection indicates where the regeneration is highest and, in theory, where the maximum or optimum catch would be obtained.

Graham (1935), after reviewing recent advances in the theory of fishing from Russell, Hjort, Jahn and Ottestad, and others, made use of the theoretical basis in assessments of North Sea roundfish and showed, via three different approaches, that a reduced fishing effort would maintain or even give higher yields. Graham's rough estimate of total production indicated that the maximum yield from these fisheries would be about 15% higher than recent levels and would be taken with 75% of the current effort. This was the first presentation in an ICES document of an estimate of the effect of changes in effort on yield.

Graham formulated his "Great Law of Fishing" which describes in general terms the relationship between yield and unrestricted fishing effort: "Because of increased fishing effort resulting from improved efficiency and addition of capital, industrial fisheries will, if left to themselves, move in a self defeating process towards a marginal state". The sigmoid curve was the more general form of this law, and in a later paper Graham (1939) demonstrated how an S-shaped curve describing the dynamic relationship between yield and effort could be derived graphically on the basis of simple biological arguments. There were problems in applying the curve to actual fisheries, but the rough estimates of the state of the stocks of North Sea groundfish presented in 1935 were again quoted. Some of Graham's general conclusions based on the model are important in that they show the attitude of a leading British scientist to fishery regulations at the time:
- There will be no permanent profit unless the rate of fishing is controlled.
- There can be only temporary prosperity until international agreement is reached to prevent the rate of fishing from increasing.

This may well be the background for British attempts to introduce effort regulation immediately after World War II. Under Graham, the Fisheries Research Laboratory in Lowestoft became a leading institution in fish population dynamics in the decades following the War.

2.7 The role of ICES as an advisory body

The resolutions for the creation of the Council were already showing that it regarded advice to the member governments as one of its primary functions. It was probably expected that such advice would follow the conclusions of the cooperative biological programmes organized under the Committees A and B. To some extent this was the case. Committee A, which dealt with stocks of fluctuating fisheries, could inform the governments and industries concerned that varying recruitment was the main cause of their good and bad years, and that predictions of the fluctuations could be made, an advance of great practical consequence.

For the "overfishing" Committee B, which mainly dealt with the effects of the high by-catches of undersized plaice, it was much more difficult to draw conclusions and submit advice, mainly because the necessary insight and understanding of the effects of fishing on stocks were not yet available. However, the conclusion of the final plaice report in 1912 was that a lower limit of 25–26 cm was extremely desirable in order to avoid the useless destruction of young plaice, and in 1913 the Council advised member governments accordingly. This initiative was lost in the war, but it seems doubtful whether it would have resulted in an effective international fishery regulation measure, as this would have needed a special international convention.

After World War I, the Council's Plaice Committee recommended the closure of inshore waters along the southern and eastern North Sea to trawling as a more practical measure for the protection of young plaice. This and other measures were discussed for a long time, as attempts were made to allocate the assumed losses of catch caused by the regulations among the different nations. This difficult and largely political task should not have been part of the remit of an advisory body.

We are therefore forced to conclude that the Council did not function as an advisory body on fisheries regulation until the special meetings in the early 1930s, especially that in 1934, on minimum landing sizes and trawl-mesh regulations. The efficacy of the latter in preventing the destruction of undersized fish was discussed. Claims that the cod-end meshes close during towing were refuted by Davies (ICES 1934) who deduced this from experiments with a special cover net that could be closed during fishing operations. Consequently there was no need for savings gear with rigid or semi-rigid meshes. Davies' contribution on mesh experiments was important, in that it incorporated a review of previous work going back to experiments with fine meshed covers as early as 1906 and included a wide-ranging discussion of the various methods that had been used: alternating hauls, covers, trouser trawls.

Iversen (ICES 1934) also contributed to the mesh discussion, confirming that the meshes remain open during towing. He estimated the size of fish which can escape through an open mesh, and he stressed the need for wider meshes in the Barents Sea since the minimum size of marketed fish taken by distance-water trawlers was higher there than in the North Sea.

The conclusions of the meetings took the form of advice to all countries fishing in the waters investigated by ICES. The recommendations were partly general, to prevent waste, to prevent the capture of young fish below the size at which they could be sold at a remunerative price, and to regulate trawl meshes in fisheries where young fish were captured. However, for the minimum size of trawl meshes and minimum landing size of fish, specific recommendations were submitted.

Britain had already in 1933 legislated for the regulation of trawl mesh size, stipulating a minimum of 70 mm in the North Sea. The ICES meeting recommended the general adoption of that mesh size for the North Sea, but 105 mm for the north-eastern area. Minimum landing sizes were adopted by various countries. It seems clear that Britain had now taken a clear lead in efforts to regulate fisheries in international waters, and the response of the governments to ICES' submission of advice on the protection of undersized fish took the form of an international conference of government representatives in London in 1937. The records of the Conference show that there was considerable resistance to mesh regulations in many delegations. France withdrew from the Conference so that agreement could be reached on a recommendation for a "Convention for the regulation of meshes of fishing nets and size limits for fish". This specified the measures advised by ICES and also included an article on the setting up of a Permanent Commission in London which would consult ICES on the need to extend or alter the provisions of the convention. The outbreak of World War II halted the process of ratification of the convention.

2.8 Summary review of the development in the period

From a general state of ignorance of the life history of the stocks and their response to exploitation at the start of the 20th century, the identity and distribution of the main stocks inhabiting the North-East Atlantic were described through a step-wise development over 40 years. Towards the end of that period some members of the scientific community had gained the first insights into the dynamics of exploited fish stocks.

During the first part of the period, investigations of the two main Norwegian stocks, the Arctic cod and the spring-spawning herring, pointed the way to general progress by the successful use of methods and through important general findings. Hjort and his group in Bergen were the leading fishery research institution in Europe until World War I, and with his stock-age composition concept, Hjort was on the verge of achieving deeper insights into the complexities of exploited fish stocks.

European fishery research then entered the doldrums for 15–20 years, caused in part by the unfortunate herring scale dispute which appears to have produced a failure of confidence in Hjort's approach and in age determination in general, at least among the British. However, a new generation of British scientists played a leading role in the development of fishery science in the 1930s, and based on Hjorts earlier findings and results of investigations of North Sea stocks, the first theories of the effects of fishing on the stocks were developed.

Over the entire 40-year period, ICES functioned as an important forum for research cooperation. However, its intended function – that of providing advice on the regulation of international fisheries – did not materialize until the early 1930s, when advice based on a precautionary principle for management was submitted, dealing with the protection of undersized fish in trawl fisheries. The effects of fishing on stocks were not regarded as urgent problems in the Norwegian fisheries in the 1920s and 1930s, with the exception of whales. However, the protection of undersized fish was regarded as important for Arctic cod too, given the growing distant-water trawl fisheries in the Barents Sea.

In other respects, Norwegian studies of cod and herring did not continue the remarkable progress of the first "golden" period. This was because the Bergen institution had lost the most important component of research from that period, the surveys performed by ocean-going research vessels. In contrast to the situation in the North Sea, the main Norwegian stocks were highly migratory fish, and survey systems, which covered their entire ranges of distribution, were necessary for full descriptions and evaluations of the stocks.

The differences in the main research interests of Norway and the North Sea countries may also have been a reason for the loss of the role of leading institution from Bergen to Lowestoft. The attention of many scientists was drawn to the North Sea groundfish stocks which were assumed to be heavily fished, and it was the experience with these stocks which led the British to take such a strong interest in theories of fishing. Their findings and the further development of fishery science represented a general advancement of knowledge and understanding which would affect all future fishery research, including that on Arctic cod and Norwegian spring-spawning herring.

REFERENCES

Broch, H. 1908. Norwegische Heringsuntersuchungen während der Jahre 1904–1906 (Norwegian herring investigations, 1904–1906). Bergens Museums Aarbog (Bergen Museum Annual Report), 1908, 1. 63 pp.

Dahl, K. 1907. The scales of the herring as a means of determining age, growth and migration. Fiskeridirektoratets Skrifter, Serie Havundersøkelser (Reports of the Norwegian Fisheries and Marine Investigations), 2(6). 26 pp.

Graham, M. 1935. Modern theory of exploiting a fishery, and application to North Sea trawling. Journal du Conseil International pour l'Exploration de la Mer, 10: 264–274.

Graham, M. 1939. The sigmoid curve and the overfishing problem. Rapports et Procès-Verbaux des Réunions du Conseil International pour l'Exploration de la Mer, 110: 15–20.

Hjort, J. 1899. Forslag til bygning af damper til norske fiskeriundersøkelser (Proposal for building a steamship for Norwegian fishery investigations). Skrivelse til Departementet for det Indre (Letter to the Ministry of Home Affairs), Kristiania. 8 pp. (In Norwegian).

Hjort, J. 1901. "Michael Sars" Første togt I Nordhavet aar 1901 ("Michael Sars". First cruise to the Northern Ocean, 1901). Aarsberetning vedkommende Norges Fiskerier for 1900 (Annual Report on Norwegian Fisheries, 1900), Bergen. 231–268 pp. (In Norwegian).

Hjort, J. 1905. Norsk Havfiskeri (Norwegian deep-sea fishing). Selskabet for de Norske Fiskeriers Fremme (The Society for Development of the Norwegian Fishing Industry), Bergen. 371 pp. (In Norwegian).

Hjort, J. 1907. Committee A. Meeting of June 11th, 1907. In C. Procès-Verbaux des Réunions des Commissions Spéciales et des Sections, 15 pp. Rapports et Procès-Verbaux des Réunions du Counseil International pour l'Exploration de la Mer, 7. 343 pp.

Hjort, J. 1908. Nogle resultater av den internasjonale havforskning (Some results of international ocean researches). Aarsberetning vedkommende Norges Fiskerier (Annual Report on Norwegian Fisheries) 1907: 351–382. Printed in Edinburgh by the Scottish Marine Laboratory, Aberdeen. 40 pp. (In Norwegian).

Hjort, J. 1909. Oversigt over norsk fiskeri- og havforskning 1900–1908 (Overview of Norwegian fishery and oceanographic research 1900–1908). Fiskeridirektoratets Skrifter, Serie Havundersøkelser (Reports of the Norwegian Fisheries and Marine Investigations), 2(1): 1–202. (In Norwegian).

Hjort, J. 1910. Report on herring investigations until January 1910. Publications de Circonstance, 53: 1–6.

Hjort, J. 1014. Fluctuations in the great fisheries of northern Europe viewed in the light of biological research. Rapports et Procès-Verbaux des Réunions du Conseil International pour l'Exploration de la Mer, 20: 228 pp.

Hjort, J. and Lea, E. 1911. Some results of the international herring investigations 1907–1911. ICES Publications de Circonstance, 61. 8–34.

Hjort, J., Jahn, G. and Ottestad, P. 1933. The optimum catch. In Essays on Populations.

Scientific Results of Marine Biological Research. The Norwegian Academy of Science, Oslo. Hvalrådets Skrifter (Reports of the Norwegian Whaling Commission), 7. 127 pp.

ICES. 1901. 2ème Conférence International pour l'Exploration de la Mer, Réunie a Kristiania. Kristiania, Norway, 1901. 51 pp.

ICES. 1903. Rapports et Procès-Verbaux des Réunions du Conseil International pour l'Exploration de la Mer, 1. 217 pp.

ICES. 1905. Rapports et Procès-Verbaux des Réunions du Conseil International pour l'Exploration de la Mer, 4. 212 pp.

ICES. 1907. Rapports et Procès-Verbaux des Réunions du Conseil International pour l'Exploration de la Mer, 7. 343 pp.

ICES. 1909a. Rapport sur les travaux de la Commission A dans la période 1902–1907. Rapports et Procès-Verbaux des Réunions du Conseil International pour l'Exploration de la Mer, 10. 841 pp.

ICES. 1909b. Rapports et Procès-Verbaux des Réunions du Conseil International pour l'Exploration de la Mer, 11. 251 pp.

ICES. 1913. Rapports et Procès-Verbaux des Réunions du Conseil International pour l'Exploration de la Mer, 19. 142 pp.

ICES. 1930a. Fluctuations in the Abundance of the Various Year-classes of Food Fishes. Reports of the proceedings of a special meeting held on April 12th 1929 in London. Rapports et Procès-Verbaux des Réunions du Conseil International pour l'Exploration de la Mer, 65. 188 pp.

ICES. 1930b. Fluctuations in the Abundance of the Various Year-classes of Food Fishes. Reports prepared by special reporters nominated by the Council and indicating the main results brought out by the papers read at the biological meeting in London in 1929. Rapports et Procès-Verbaux des Réunions du Conseil International pour l'Exploration de la Mer, 68. 115 pp.

ICES. 1932. The Effect upon the Stock of Fish of the Capture of Undersized Fish. Reports of the proceedings of a special meeting held on June 24th 1932, in Copenhagen. Rapports et Procès-Verbaux des Réunions du Conseil International pour l'Exploration de la Mer, 80. 85 pp.

ICES. 1934. Size-limits for Fish and Regulations of the Meshes of Fishing Nets. Reports of the proceedings of the special biological meeting held on June 4th and 8th 1934, in Copenhagen. Rapports et Procès-Verbaux des Réunions du Conseil International pour l'Exploration de la Mer, 90. 61 pp.

Lea, E. 1910. On the methods used in the herring investigations. ICES Publications de Circonstance, 53: 7–174.

Lea, E. 1929. The herring scale as a certificate of origin. Its applicability to race investigations. Rapports et Procès-Verbaux des Réunions du Conseil International pour l'Exploration de la Mer, 54: 21–34.

Rollefsen, G. 1933. The otoliths of the cod. Preliminary Report, Fiskeridirektoratets Skrifter, Serie Havundersøkelser (Reports of the Norwegian Fisheries and Marine Investigations), 4(3): 1–18.

Rollefsen, G. 1966. Norwegian fisheries research. Fiskeridirektoratets Skrifter, Serie Havundersøkelser (Reports of the Norwegian Fisheries and Marine Investigations), 14(1): 1–36.

Russell, E.S. 1931. Some theoretical considerations on the "overfishing" problem. Journal du Conseil International pour l'Exploration de la Mer, 6: 3–20.

Ruud, J.T. 1971. Einar Lea 1887–1969. Journal du Conseil International pour l'Exploration de la Mer, 33: 303–307.

Schwach, V. 2002. Internationalist and Norwegian at the same time: Johan Hjort and ICES. ICES Marine Science symposia, 215: 39–44.

Sinclair, M., Solemdal, P. 1987. Development of population thinking in fisheries research between 1878 and 1930. ICES CM 1987/L:11. 54 pp.

Went, A. 1972. Seventy Years Agrowing. A History of the International Council for the Exploration of the Sea, 1902–1972. Rapports et Procès-Verbaux des Réunions du Conseil International pour l'Exploration de la Mer, 165. 252 pp.

~ CHAPTER 3

Technological developments in Norwegian fisheries

Steinar Olsen

At the beginning of the 20th century, traditional fishing gears such as hook and line, longlines and gillnets, were generally used in the cod fisheries, and by and large this situation persisted until World War II. British vessels, and later also some from other nations, fished with otter trawls in the Barents Sea, first for plaice, later mostly for cod. Norwegian fisheries for cod and other groundfish gradually also expanded to offshore banks and to the Bear Island and Spitsbergen fishing grounds. From 1926 on, a few small, Norwegian motor trawlers also joined the offshore cod fisheries, as did a handful of steam trawlers from the early 1930s. Danish seining, introduced for catching plaice early in the century, was also used for cod fishing from the mid-1920s on.

Although by the start of the 20th century, steam-engined vessels had already been operating for some time in the Norwegian fisheries, they never came to play a similar role as in Britain and other fishing nations around the North Sea. In Norway, the major mechanization break-through, not only with regard to vessel propulsion, but subsequently also to gear and catch handling, came with

Figure 3.1 *Traditional open, 8-oared rowing/sailing boat converted to a decked fishing vessel with motor (see Johansen, 1999).*

the rapid motorization of the fishing fleet, which took off during the first decade of the century (Figure 3.1).

This development was financially supported by grants given directly to individual fishermen by the NGO "Selskabet for de Norske Fiskeriers Fremme", which thereby came to play a significant role in promoting the change. During the first half of the 20th century gradual improvements and innovations in fish-capture technology were mainly generated by the fishing industry itself. After World War II, however, government funds were provided for goal-oriented research and activities to test new inventions, and these funds played a significant role in the technological development of the Norwegian fisheries.

Mechanized gear handling with power-driven gurdies for hauling longlines, gillnets, seine and other ropes, etc., expanded quickly soon after the introduction of motor propulsion. Another important novelty that greatly improved operations and working conditions on the fishing vessels was electric light. According to fishing skipper Johannes Olsen (personal communication), longline and gillnet fishing on the coast of Finnmark during the darkest and stormiest winter months had been extremely difficult and scarcely profitable before reliable electric buoy lights became available in the 1920s.

The herring fisheries were traditionally pursued with gillnets set on the bottom. From the end of the 19th century onwards, herring gillnets were also operated as driftnets near the surface; this also facilitated offshore fishing. Large shore (beach) seines operated from open boats were used to impound near-shore herring schools. As early as 1876, an American type purse seine was unsuccessfully tested in the herring fishery in Rogaland, and in the Oslofjord area, purse seines of a type used on the west coast of Sweden were introduced before the turn of the century. However, the breakthrough in the Norwegian herring fisheries for this versatile and highly efficient fishing method did not come until an American-made purse seine had proved to be successful at Iceland and on the west coast of Norway in 1904.

Just as in the cod fisheries, new innovations and better operational practices were tested and introduced in the herring fisheries in the course of the years. Thus, herring gillnet hauling soon became mechanized, and various methods and systems were later introduced for shaking the herring from the nets once they had been hauled on board. The purse seine, at first carried in a single rowed net-boat, soon became operated using a pair of seine boats, dories, each carrying half the seine. This made it easier to rapidly surround schools of herring. Pursing the net, which had been done by hand-operated gurdies, also became a power-driven process when the dories were motorized.

By their very nature, both beach and purse seine fishing are aimed types of fish capture, and to locate the herring schools and determine their depth distribution, the fishing master, the "Bas", was being rowed around in a small boat with a submerged lead weight attached to a brass wire below the boat (Figure 3.2). This method of locating fish concentrations was also utilized in purse seine fishing for other types of fish, such as large cod in some Finnmark fjords before the end of World War II. However, when cod purse seining was permitted during the Lofoten winter fisheries in 1949–59, echo-sounders had been introduced.

As early as the 1930s, a few purse seine vessels had installed echo-sounders to better locate fishing grounds with herring schools, but hydroacoustic instru-

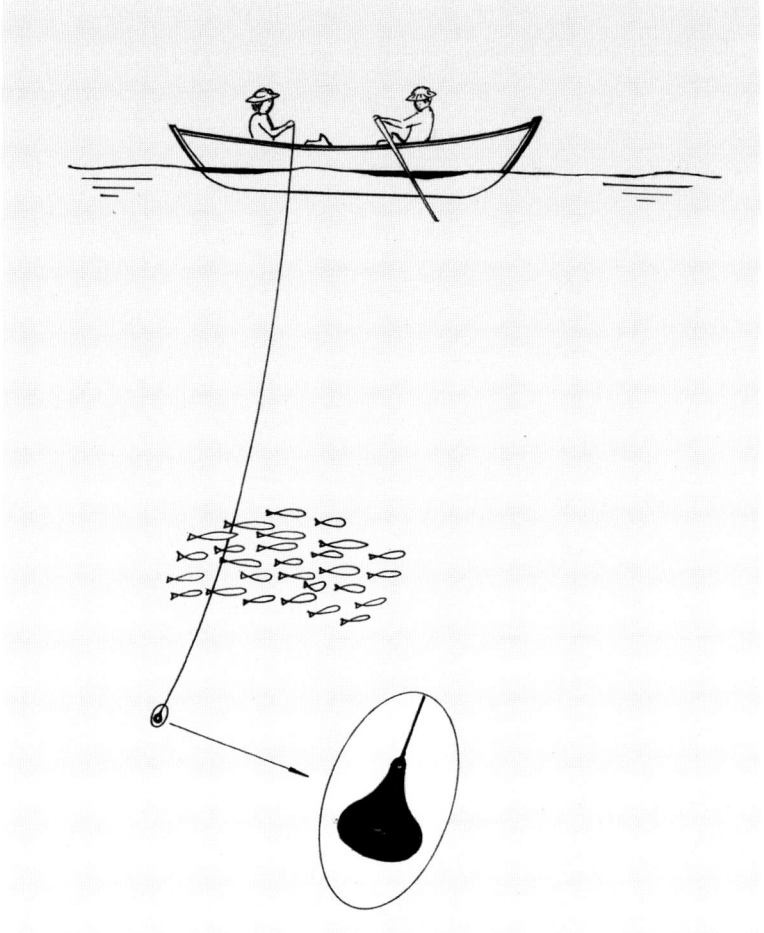

Figure 3.2 The "Bas" locating fish concentrations by lead and wire.

ments as direct aids in purse seine operations did not come into widespread use until after World War II.

The postwar period has witnessed an almost continuous revolution in fish-capture technology. Rebuilding and modernizing the fishing fleet and the gears and methods employed was given high priority by the Norwegian government after the War. Thus, foreign exchange was released and ear-marked for imports of new strong, non-rotting synthetic fibres suitable for fishing gears, as well as for importing a limited number of echo-sounders to be distributed to selected fishing vessels in each coastal county. The National Defence Research Institute put high priority on the development of an indigenous echo-sounder, and as early as 1947, SIMRAD, a manufacturer of electronic equipment, was licensed to develop this further for use in the fisheries and to commence production of high-performance hydroacoustic instruments for use in commercial fisheries and fisheries research. This launched a to date unbroken series of cooperative activities of mutual benefit to IMR and SIMRAD (Sogner 1997).

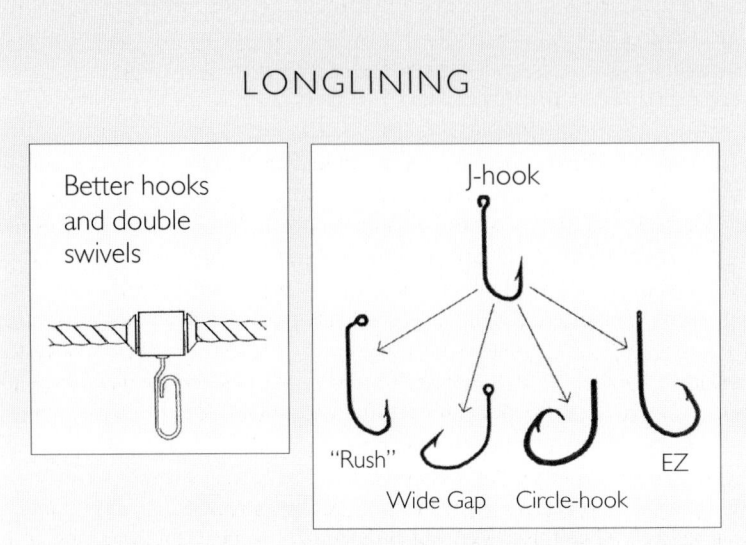

Figure 3.3 *More efficient hooks, swivel attachments of snoods and other longline innovations were rapidly adopted.*

In the early 1950s, gillnets made of synthetic fibres, mainly nylon, were tried in the cod fisheries. These trebled or quadrupled catching efficiency (Sætersdal 1959), and by the mid-50s, only synthetic fibres were used for netting as well as for head- and ground lines in the gillnet fisheries. Plastic rings were then adopted for floats and iron rings for sinkers, although these subsequently replaced by ground lines with a lead kernel. The depth of the nets was greatly increased in bottom fisheries.

In longline fishing, new, less visible, synthetic line materials (e.g. monofilament), swivel attachment of snoods to the mainline, and new types of hook have all greatly enhanced catching efficiency in all Norwegian long line fisheries (Figure 3.3).

The invention and introduction of mechanized longline handline and baiting systems (autoline), especially in larger offshore longline vessels, have greatly contributed to the conversion of this old, traditional method of fishing into a high-tech, industrial fishery (Bjordal and Løkkeborg 1996).

In the coastal fisheries with smaller vessels (generally 8 to 14 m OAL) similar improvements in gillnet and longline fish capture technology were rapidly introduced. In addition major advances in handline (jigging, trolling) fishing technology since World War II have greatly impacted the development of this segment of the fishing fleet, and this development has been entirely industry-driven. These old methods of fishing have been developed into high-efficient, profitable fisheries carried out from polyester/glass-fibre diesel-powered boats equipped with radar, GPS and other modern electronic navigation instruments, remote steering and engine controls on deck, and with one or more high-tech automatic jigging and/or trolling machines controlled by a data processor that offers a range of operation programmes (Figure 3.4). The fishing gear is constructed from monofilament nylon rigged with the appropriate lures for the particular type of fishing involved. In Scandinavia, these high-efficiency fish capture machines are mostly manned by a single man to catch whitefish on the bottom as well as pelagic species such as mackerel in midwater. When conditions are suitable for hook and line fishing, the average net earnings per man-day can scarcely be matched in any other fishery.

Danish (Scottish) seining became a common fishing method in the Lofoten large cod (skrei) fisheries from the 1950s onward. New developments in gear and methods of operation were gradually being introduced, e.g. rope drums and power blocks for easy rope and net handling. Lead rope droppers, ground rope skirts and ropes with steel inlays instead of lead made Danish seining for round fish possible in localities that were previously considered unsuitable for seine fishing. Many coastal gillnet vessels, as well as some longliners were converted to Danish seiners. By 1992, these numbered 300, as against 100 only four years earlier, and recently some autoliners have also converted to Danish seining.

The introduction of new strong, non-rotting synthetic fibres also revolutionized purse seine fishing. Very large nets that were also suitable for efficient high-seas fishing, could now be made, which previously had been inconceivable because nets of the required size would burst under their own weight in operation if made from natural fibres. The dories were dispensed with, and the purse seine net was now operated directly from the main vessel. To handle such big and heavy purse seines, the American Puretic Power Block was

Figure 3.4
Modern coastal multi-purpose fishing boat.

Table 3.1 *Chronology of major technological developments in fisheries.*

Period	Development
1900–1920	Vessel motorization
	Mechanized gear handling
	Electric lighting
	Herring purse seining
1920–1940	Cod longline and gillnet fishing on offshore banks
	Start of demersal offshore trawl fisheries
1945–1965	Synthetic Fibres
	Hydroacoustic fish finding
	Highseas "purse seine revolution"
1970–1985	Pelagic trawl development
1975→	Evolution and introduction of the modern high-technology glass-fibre small coastal fishing craft
1985–1995	Development of Danish seining technology facilitating roundfish seining in locations with rough bottom

introduced in 1957. Soon, Norwegian-made net winches came on the market, the TRIPLEX becoming the most popular of these. The design of purse seines had to be modified to allow them to be hauled with a net winch (Hamre and Nakken 1971), and several modifications have been introduced in the course of the years (Figure 3.5).

The collapse of the stock of spring-spawning herring around 1970 raised interest in another major oceanic resource of pelagic fish, the blue whiting. Efficient large midwater trawls were developed with active input from relevant industries and research institutes in the Faroes, Iceland and Norway, and a large high-seas fishery for blue whiting developed in the latter half of the 1970s. The technology has subsequently been further advanced and employed in the herring and mackerel fisheries.

The rapid and important developments in Norwegian and world fisheries were largely facilitated, and often in fact started, by the adaptation of new technological innovations and improvements, not least those generated by the vastly intensified research and development efforts for specific military objectives during World War II. Thus, while echo-sounders had been installed in some few fishing vessels in the early 1930s, it was the significant progress in hydroacoustic technology for anti-submarine warfare that found widespread application in new and better instruments for fish detection and location soon after the War. Not only did the new efficient echo-sounders (and later the horizontal ranging sonars) vastly improve the ability to detect mid- and deep-water concentrations of fish, but they also became indispensable aids in the actual operation of fishing gear, as well as greatly extending search range and thus opening up the offshore grounds for rapid and efficient fish location. Similary, the rapid developments of improved electronic navigation and telecommunication technology also quickly became standard in the industrialized fisheries. Equally important was the emergence on the market shortly after World War

II of strong, non-rotting, synthetic fibres that facilitated the construction of much larger and stronger fishing nets than had previously been conceivable. Simultaneously, new developments, especially in hydraulics, were adapted and employed in novel versatile deck machinery designed to handle and operate the much larger and heavier fishing gears that were coming into use. Meanwhile, freezing and cooling systems, such as RSW (refrigerated sea water), made it possible for fishing vessels to preserve their catch in good condition during long voyages. In conclusion, therefore, it was largely the almost continuous flow of technological advancements that governed and intensified developments in the world's fisheries during the second half of the 20th century.

Parallel to the spectacular rate of technological innovation in fishing gear technology and its rapid adoption, technology has also advanced greatly in all sizes and types of fishing vessel. In fact, the vast development of world fisheries would have been inconceivable without the parallel – often paired – evolution of better fishing craft and more efficient fish capture gear and methods. In Norway the development in fishing vessel design was largely industry-driven, but the naval architecture research environment in Trondheim has also played a significant role.

Figure 3.5
Modern Norwegian purse seiner.

REFERENCES

Bjordal, Å., Løkkeborg, S. 1996. Longlining. Fishing News Books Ltd. Oxford.

Hamre, J., Nakken, O. 1971. Technological Aspects of the Modern Norwegian Purse Seine Fisheries. In: Modern Fishing Gear of the World, Vol. 3 (ed. H. Kristjonsson). Fishing News (Books) Ltd. Oxford.

Johansen, K.E. 1999. Moderniseringa av norsk fiske ca. 1880–1920. Årbok 1999, Norges Fiskerimuseum. ISBN 82-92257-00-4, Bergen 2000 (In Norwegian).

Sogner, K. 1997. God på bunnen. SIMRAD-virksomheten 1947–1997, Oslo (Novus Forlag) (In Norwegian).

Sætersdal, G. 1959. On the fishing power of nylon gill nets. In: Modern Fishing Gear of the World, Vol 1 (ed. H. Kristjonsson). Fishing News (Books) Ltd. London, pp. 161–163.

CHAPTER 4

Norwegian spring-spawning herring: history of fisheries, biology and stock assessment

Olav Dragesund, Ole Johan Østvedt and Reidar Toresen

4.1 Introduction

Compared with other herring stocks, Norwegian spring-spawning herring are distributed over wider areas, perform more extensive migrations, and are characterized by a longer life cycle and a greater stock abundance. Figure 4.1 shows the general potential geographical distribution of this stock. This is the distribution the stock is known to have had during the century of scientific studies of its distribution. Figure 4.2 shows the geographical distribution and main migration patterns of adults since 1950. This was a period of great fluctuations in stock size and significant changes in distribution patterns. The Norwegian spring-spawning herring live in a more productive zone of the oceans and have a better food supply than many other herring stocks in the Atlantic and Pacific Oceans. The habitat of Norwegian spring-spawning herring is influenced by warm Atlantic water penetrating the Norwegian Sea and the Barents Sea as well as adjacent Arctic water from the Polar basin. Extensive drift migration of larvae and fry enables the herring to colonise large areas that are suitable for their

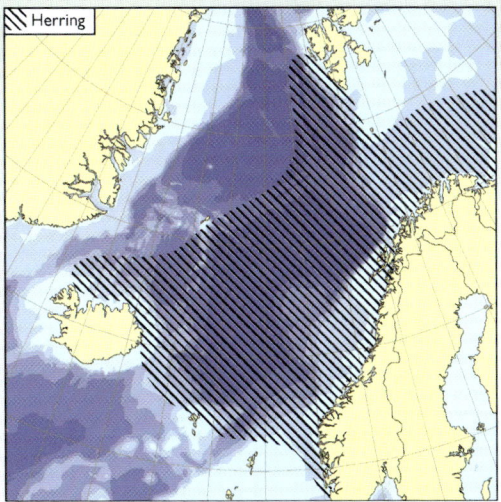

Figure 4.1
Approximate total known distribution area of Norwegian spring-spawning herring (Holst et al. 2004).

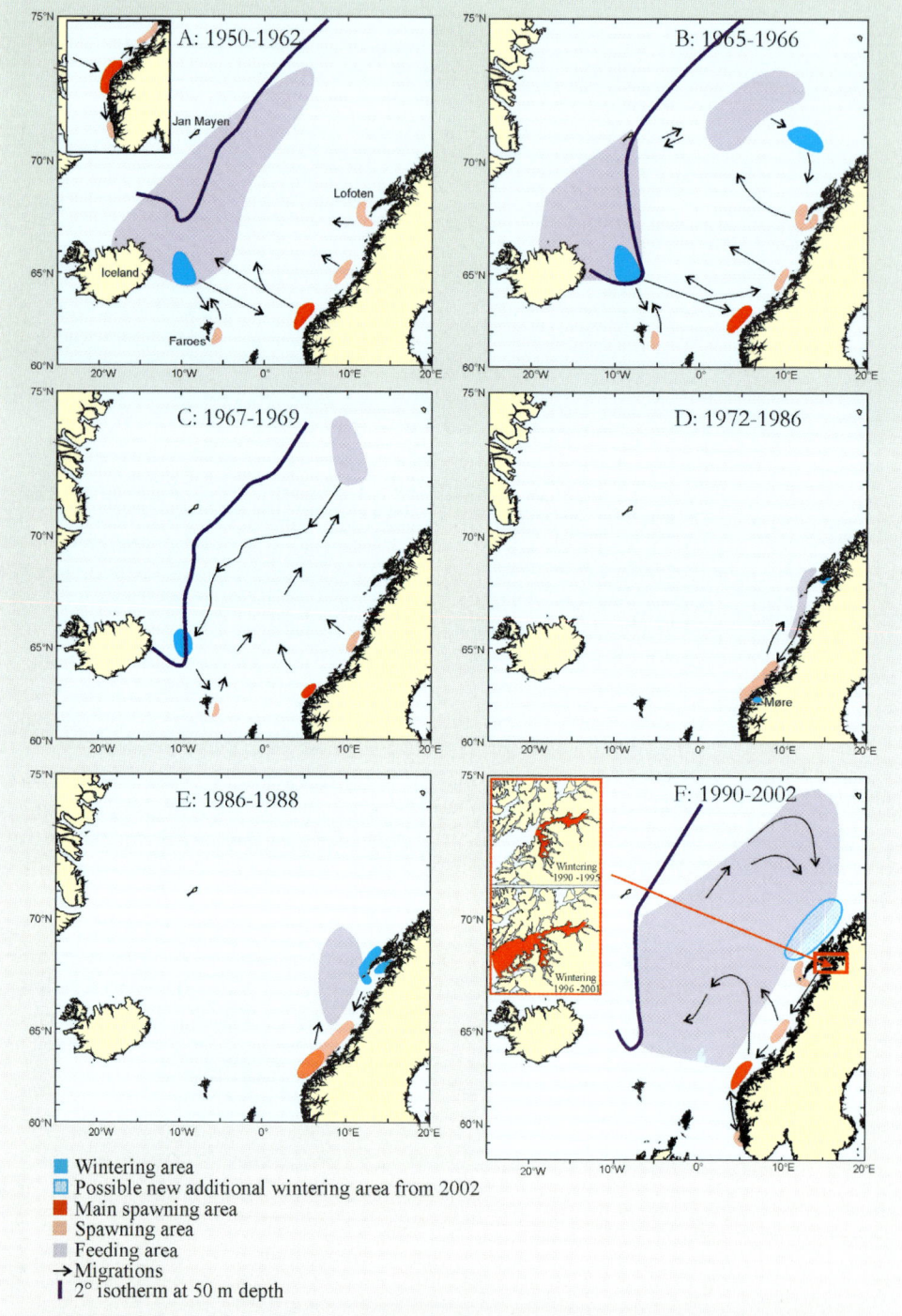

Figure 4.2 *Main migration patterns of the Norwegian spring-spawning adult herring during the period 1950–2002. (Holst et al. 2004).*

existence. The adult stock consists of 12–15 age groups, and sexual maturation takes place within an age range of 3–9 years, depending on abundance. Although Norwegian spring-spawning herring are long-lived, the spasmodic pattern of recruitment means that the stock fluctuates on a long-term basis.

In the course of the centuries, Norwegian fishermen have experienced long periods when herring were easily available along the Norwegian coast. In between these periods there was scarcity, not only of herring, but also of fish and other organisms that habitually fed on this rich pelagic resource. The decrease in abundance through the 1950s has been explained by less favourable conditions for survival of herring larvae, but in the 1960s the stock collapsed because the fishery expanded rapidly at the same time as the stock was in a low productive phase. However, it is likely that the large fluctuations in availability of herring in earlier periods reflect long-term variability in abundance and that this is a natural phenomenon for this stock (Dragesund *et al.* 1997).

Toresen and Østvedt (2000) analysed fluctuations in stock size throughout the 20th century and found two periods of high productivity, one in the first half of the century and another in its final decade. The development of the spawning stock biomass in the course of the 20th century is shown in Figure 4.3. The stock was in a rather poor condition by the end of the 19th century even though exploitation was very low at the time. The estimated spawning stock biomass (SSB) in 1907 was 1.6 million metric tonnes. The stock increased in abundance due to good recruitment, reaching high levels of more than 10 million tonnes in 1928–1937 and 1942–1957, with the highest recorded estimate of 16.2 million tonnes in 1945. It then decreased and collapsed in the late 1960s. Throughout the 1970s, the stock was estimated to be at very low levels, but slowly recovered again through the 1980s. At that point, the increase in biomass accelerated due to the recruitment of a few rich year-classes, those of 1983, 1991 and 1992, attaining levels of 8–10 million tonnes by the late 1990s. Toresen and Østvedt (2000) found these long-term stock size fluctuations to be natural, as they were highly correlated with the climate conditions in those parts of the ocean where the herring live (Figure 4.3).

The collapse of the stock in the 1960s was due to heavy fisheries, whereas the downward trend in the spawning stock biomass in the 1950s, prior to the collapse, was found to be driven by the influence of environmental conditions at recruitment levels. In other words, it is believed that through the 20th century we have experienced two herring periods, one starting by the turn of the century

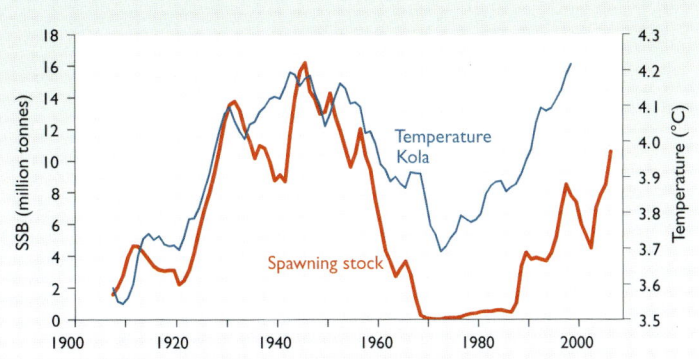

Figure 4.3
Fluctuations of spawning stock biomass of Norwegian spring-spawning herring (red line) and mean annual (moving average over 19 years) temperature (blue line) at the Kola section (updated after Toresen and Østvedt 2000).

(around 1900) with a period of high abundance through the 1930s and 40s that ended in the early 1970s. The second period started in the 1980s, with a rising trend in biomass through the final decades of the 20th century. The peak of the second period has not yet been reached. The start of the two periods has certain similarities. They both started with the recruitment of a very strong year-class while the spawning stock was at a low level of abundance, the 1904 year-class for the first period and the 1983 year-class for the second. In both periods, these two year-classes dominated the fishery for a number of years.

Fishermen have experienced these fluctuations in stock size as variations in availability. During the 1960s, when the stock collapsed, there was a fundamental development in fishing gear and technology (see Chapter 3) which helped fishermen to exploit the stock even while it was falling so dramatically in abundance. On the other hand, through this period of decreasing stocks, the scientists developed stock monitoring and analytical assessment methods as a basis for fishery management. The collapse in the stock also led to a firmer fisheries management based on results from the assessments. It is therefore natural to divide the history of research on Norwegian spring-spawning herring through the 20th century into two parts; the first describing the period from about 1900 to about 1970, and the second the period after the collapse (from 1970–2000). In the period before the collapse, the fishermen saw the role of the scientists as supervisors, guiding them to the largest concentrations and thus helping them to take bigger catches. In the early 1950s, only the research vessel was equipped with sonar. After the collapse, and in the stock rebuilding phase during the 1970s and 80s, the role of the scientists in relation to the fisheries changed. Their guiding role for the fishermen was not continued and the scientists now provided fisheries managers with information based on a new set of criteria, to reduce the fishery sufficiently to enable the stock to be rebuilt for sustainable fishery management. The scientists also put greater emphasis on research for the development of better abundance estimation methods and assessment of fish stocks.

It must be kept in mind that, at the beginning of the 20th century, our knowledge of the biology of the herring stocks in the Northeast Atlantic was very limited. Neither the fishermen nor the scientists dealing with the biology of fish stocks knew about the difference in distribution or migrations of the various herring stocks in the region; the Norwegian spring-spawning herring, the Icelandic spring- and summer-spawning herring, or the herring stocks in the North Sea. Nor was there any knowledge of the reasons for the fluctuations in the yield of the fisheries. However, Hjort developed very early the idea of fluctuations in the year-class strength as the main reason for the great fluctuations in fish stock abundance (see Chapter 2). These theories formed a basis for the research on fish stock biology and dynamics carried out by subsequent generations of scientists, and led to the analytical assessments and thorough knowledge of fish stock dynamics of today.

The aim of the present chapter is to provide a synthesis of:
- the development of the herring fisheries in two periods, from about 1900 and the following 70 years, and from about 1970–2000
- the distribution and migratory pattern of the herring in the same periods
- the progress in knowledge of herring population biology
- the development of knowledge of stock abundance and management

4.2 Yield of the herring through the 20th century

The Norwegian spring-spawning herring fisheries are based on fish representing various stages in the life history and are best treated as three main branches; the winter herring fishery on the spawning grounds on the Norwegian west coast, the summer adult herring fishery during the feeding period off the north and east coast of Iceland, and in recent years, in the Norwegian Sea, and the young and adolescent herring fishery (0, I and II–III group) on the nursery grounds in the Norwegian fjords and coastal waters. Before the collapse of the stock in the 1960s, the fishermen characterized the herring caught during the winter herring fishery as "large herring" before the herring spawned, and "spring herring" while and after they had spawned. The reason for this is that the quality of the fish fell after spawning. For practical reasons, a specific date was decided on each year, after which the herring was called "spring herring", and this was often the first of February.

Landings of Norwegian spring-spawning herring throughout the period 1907–1997 are shown in Figure 4.4. In the very early part of the 20th century, total landings were about 200 000 metric tonnes, but soon rose to a level of about 250 000–300 000 metric tonnes. The total quantity increased somewhat during the first 25 years of the century, to a level of about 400 000 tonnes by 1926. During the following five years, total landings continued to rise, to a level of about 700 000 tonnes. This level was held until the late 1940s. During the 1950s, the fisheries expanded, and the total catch reached a level of about 1.3 million tonnes. In the 1960s, the exploitation continued to increase, and in 1966, the total amount landed was nearly 2 million tonnes. Thereafter, the catch dropped drastically to less than 100 000 tonnes in 1969. The low level continued throughout the 1970s. There was a total ban on the herring fishery in 1973–1975, but even with the ban, small amounts of herring were landed, but through 1972–1982 less than 20 000 tonnes were taken annually except in 1977, when 23 000 tonnes were landed. From 1985 on, the fishery expanded once again, but remained at a fairly low level of less than 200 000 tonnes until 1994. Through the 1990s, there was a new increase in the landed quantity, and in 1994 the total amount was 480 000 tonnes. In 1995, 905 000 tonnes were landed, and in 1996–1998, landings exeeded 1.2 million tonnes.

Figure 4.4
Landings of Norwegian spring-spawning herring throughout the period 1907–1997. Blue line = total landings, red line = landings of young herring (Toresen and Østvedt 2000).

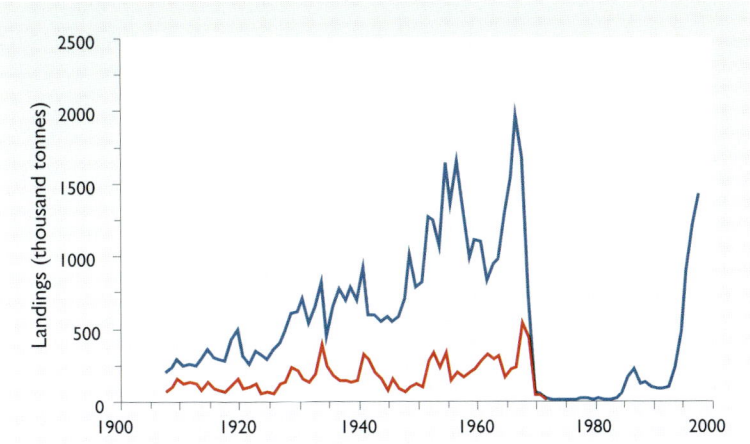

4.3 A brief history of the fisheries and their development in the period 1900–1970

Fishing gear and technique

Before 1900, bottom gillnets and beach seines were the most important fishing gears for winter herring in Norway. These gears were set close to the shore, often from small, open boats, and hauled ashore (seines) or into small boats (nets).

At the end of the 19th century, there was a development in fishing gear. During the final decades of the century, the yield of the herring fisheries had been poor because the herring were not available close to shore. The fishermen therefore moved further offshore where herring had been observed, but this required other types of gear that were fit for operation in open waters. Drift-net fishing for herring in Norway had been tried several times through the 19th century without much success. However, at the end of the 19th century, the drift-net fishery developed and was operated offshore, catching large herring and gradually becoming more important (Vollan 1971). The fishery for large herring was very successful because the fish were of better quality than the spring herring. A drift-net fishery also started off Iceland in the early years of the 20th century. Drift-nets were also used in the fishery for adolescent herring in late summer and autumn off the coast of northern Norway. The drift-net fisheries opened a "new world" to Norwegian herring fishermen. They could "meet" the herring on its way to the spawning grounds and the period during which they could fish high-quality pre-spawning large herring was considerably longer. However, bottom nets were still used quite regularly, and about equal proportions of the herring landings during the winter fishery came from the two kinds of nets until about 1950. From about 1950, use of the bottom nets decreased drastically. Beach seines continued in use far into the 20th century, but were of minor importance after 1950.

Another important type of gear that began to be operated by fishermen during the early years of the 20th century was purse seine nets. This gear was mainly used for catching spring herring in the fjords or near the coast. The seine was set from two smaller boats (dories), each carrying one half of the net. During the first 30 years of the century, the dories were rowed during the fishing operation. To operate the gear efficiently, the fishermen were therefore dependent on fair-weather conditions, or had to operate in inshore waters. At the beginning of the 1930s, the dories were equipped with motors. This made the work easier, but the operation of the gear was still dependent on fair-weather conditions, as the nets were still "dried", i.e. hauled, by man-power in the dories (Vollan 1971). The purse seine fishery was therefore often limited by bad weather, while the drift-net fishery could operate under rougher conditions. Figure 4.5 shows the operation of the purse seine before the introduction of the power block in the early 1960s.

The course of the winter herring fishery in the period from 1900–1950 was: a drift-net fishery offshore when the large herring arrived at the northwest coast of Norway, and when the herring started to spawn, the fishermen changed to bottom-set gill-nets on the spawning grounds. The beach seine was also an important gear in inshore waters until the 1940s.

A purse seine fishery in coastal waters developed through the 1920s and 30s, mainly for spring herring. Later in the period (in the 1940s and 50s) the fishery expanded from an inshore fishery, where the fishermen waited for the herring,

Figure 4.5
The operation of the purse seine before the introduction of the power block in the early 1960s. MS "Helganes" of Vedavågen operating the net in 1955. Photo: Vermund Rasmussen

to a fishery where the fishermen searched out the resources, not only inshore, but also in more open waters (Vollan 1971).

During the 1950s, the fishing fleet started to use echo-sounders and sonar to search for the herring (see Chapter 3). The introduction of ring nets and net-hauling systems on board the herring purse seiners in the early 1960s, enabled them to operate nets which were much bigger. Furthermore, the nets were made of a new, lighter material, nylon. With the power block, the fishermen were also able to handle much larger catches. The power block allowed the fleet to operate in more open waters as they were no longer dependent on the small dories. This development increased the efficiency of the herring fishery considerably.

The herring fishery during winter (large and spring herring)
The large and spring herring are all sexually mature. The large herring are adult fish with hard roe and milt (pre-spawning stage), whereas the spring herring are at the point of spawning (spawning stage) and spent herring. When in the course of the 20th century the fishermen gradually became better at catching the large herring, they began to distinguish between the large and the spring herring fisheries. However, since there is no sharp boundary between the two fisheries, the term "winter herring fishery" has also been used.

As pointed out in Chapter 1, the long-term history of this fishery has shown a marked periodicity in yield, with periods of high catches alternating with periods of great scarcity. The distribution of herring catches in the

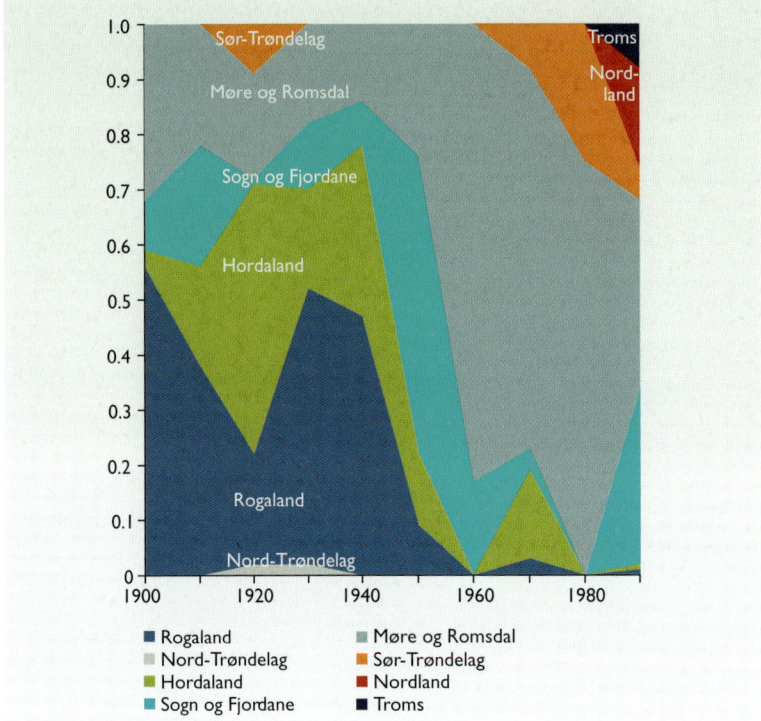

Figure 4.6
The distribution of the catches of herring in different counties during the winter season in different periods, 1900–1990.

winter season in various periods is illustrated in Figure 4.6. In practice, the winter herring fishery did not exist from 1875 to 1882. However, in the late 1880s, the herring again gradually began to visit the spawning grounds off the southwest coast of Norway. As in the earlier herring period, lasting from 1807 to 1874, the fishery in the first part of the new period was limited to the old fishing grounds off the southwest coast of Norway, between Stavanger and Haugesund (Figure 4.6).

Further north along the coast, north of Bergen, there was virtually no fishery until the turn of the century, apart from an occasional fishery for large herring. In September 1886, there was a major immigration of pre-spawning herring between Bergen and Stad. The time of arrival and the kind of herring were unexpected by the fishermen, who had most experience of the spring herring fishery. It is likely that these herring were of the same kind as those that had appeared off northern Norway at the end of the previous period (1871–74). A rich fishery for this herring took place throughout the autumn and winter of 1886/87. However, the winter herring fishery was almost negligible during the following years north of Bergen until 1896, when a rich herring fishery took place in November and December. In the autumn and winter of 1897/98, the large herring also appeared further north (Sunnmøre), and the fishery lasted from September to January. This indicates that in between the herring periods, when abundance probably was low, the stock wintered on the coast or in near-offshore areas. During this period, some herring were available for fishing on different parts of the coast from autumn to spring.

Since the turn of the century, the large herring appeared regularly every year north of Stad, mainly in offshore areas, and did not come close to the shore.

Thus, after 1900 and for the following 50 years, herring were caught during winter from the south-western coast of Norway to the Møre–Trøndelag area, and in some years even further north (Nordland). During this period, the winter herring fishery had two seasons, one based on large herring (pre-spawning stage) caught in an area south and north of Stad, and the other on spring herring (spawning stage), mainly caught south of Bergen, although some spring herring were also caught in the large herring district.

The geographical distribution of the spring herring fishery on the southwest coast of Norway was relatively constant up to 1950, whereas the large herring fishery in the same period fluctuated very much. When the large herring, after several years of absence (since 1871–74 in northern Norway), reappeared in coastal waters at the beginning of the last century, a fishery started on the Møre coast and Trøndelag, indicating that this area was a core region for the spawning stock. Between 1920 and 1950 the most important large herring fishery took place somewhat further south (between Bergen and Stad), suggesting that when the stock is large, the herring expand their spawning grounds along the coast.

The time of the first catches of large herring in coastal waters has also fluctuated, as indicated in Figure 4.7. The herring appeared as early as September during the first years (1900–1905) off the Trøndelag/Møre coast. Gradually, the large herring began to arrive later in the autumn, and by the end of the 1920s, somewhat further south around Stad. During the years 1920 to 1940, the herring arrived in December/January and in the mid-1940s in the second half of January. There seems to be a relationship between the time of arrival of the large herring and the size of the spawning stock. High stock levels result in expansion of the feeding area – and probably this in turn will influence the time and place of arrival at the coast. When the stock is at a low level, the herring appear early in the autumn, while if the stock is large, their arrival on the coast is late (January–February). This behaviour is probably related to the location of the wintering of the stock. When abundance is low, the herring tend to winter closer to the coast and in fjords along the Norwegian coast, while if the stock is large, it will winter in the Norwegian Sea.

Figure 4.7
The time of the first catches of large herring in the coastal waters (Aasen and Dragesund, personally delivered).

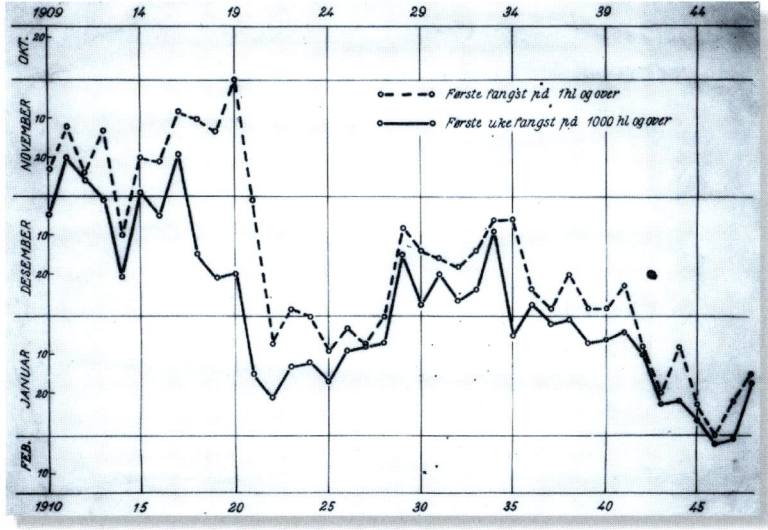

The most important fishing grounds are the Møre coast, the area between Stad and Bergen, and off the southwest coast (Stavanger–Haugesund). Throughout the 1930s, the fishing grounds south of Bergen (near Haugesund) were at least as important as the Møre coast. The grounds off the southwest coast seemed to be more important in periods when abundance was high. When stocks fell in the 1950s, the herring gradually disappeared from the spawning grounds off southwest Norway, and 1959 was the last year these fish were caught on these grounds in this former herring period.

The stock collapsed in the mid-1960s, and by 1968–1971, catches in the winter herring fishery were negligible compared to those in earlier years.

Summer and autumn herring fisheries off Iceland and in the Norwegian Sea
Another important adult herring fishery takes place on the feeding grounds in the Norwegian Sea. The fishery was located for a long time off north and northeast Iceland. Norwegian herring fisheries off Iceland started with some small attempts in the 1860s, but developed rather rapidly so that by the 1880s, some 300–400 vessels with approximately 3000–4000 men were taking part. In the 19th century the fisheries were carried out close to shore and in the fjords, because of the primitive, small open fishing boats that were used. The gear was bottom-set nets (mesh size 2–3 cm) and small beach seines (100–300 m length and 25–35 m depth).

By the turn of the century, gear and fishing techniques improved, and also off Iceland the fishermen started to use drift-nets and purse seines, so that a larger summer herring fishery on adult herring (Nordurlandssild) gradually developed on the coast of North Iceland. During the first two decades, most of the herring were caught by Norwegian fishing vessels. It was not until after World War I that most of the catch was taken by Icelandic vessels (Jakobsson 1980).

However, fishing was also carried out in the fjords, and frequently started in and outside Isafjorddypet and around Cape North. Later on, the fisheries were carried out further offshore, northeast of the coast, mainly during the summer/early autumn.

The exploitation of herring around Iceland was still at a comparatively low level at the beginning of the 20th century. In the first two decades, the annual total yield of what is believed to be Norwegian spring-spawning herring was at a level of about 20 000 tonnes. In the late 1920s, annual catches increased to 60–70 000 tonnes for a 10-year period. This was mainly because of the development of the fish-meal industry, which made the market for herring substantially larger than previously. Thereafter the yield increased further, but varied substantially, for about 20 years.

During the 1960s, the exploitation of herring accelerated because new technical advances improved the purse seining system in this fishery, first in Iceland and thereafter in Norway. The new technique made it possible to extend fishing seawards, and in the late 1950s and throughout the 1960s, the fishery took place far offshore in the Norwegian Sea, and the season lasted until October–November (Nakken 1998). The peak in the landings in the mid 1960s illustrates this feature (Figure 4.4).

Another reason for the increase in the exploitation was the development of the Soviet drift-net fishery in the Norwegian Sea in the 1950s (Marty and Fedorov 1963). Initially, this fishery took place during the summer, exploiting the areas in the Norwegian Sea between Spitsbergen–Jan Mayen and Iceland. In the 1950s, an autumn-winter fishery started along the migratory routes of the pre-spawning herring, from the overwintering areas east of Iceland towards the spawning grounds at the Norwegian coast. The Soviet drift-net fishing technique was also improved. During the late 1950s, the vessels taking part in the fishery were gradually equipped with echo-sounders, which made the sets more productive.

The drastic change in the migratory pattern of the adult herring stock during the 1960s greatly influenced the location of the summer and autumn fisheries. Off Iceland it became negligible and most of the fishery took place off the shelf west of Bear Island–Spitsbergen. In 1962–1966, there was a sharp increase in the total yield of this fishery and in 1967 about 1 million tonnes were landed, after which the fishery collapsed (Figure 4.4).

The abrupt change in availability of herring in Icelandic waters in the late 1960s was a clear sign of the change in the abundance of the stock, leading to a more limited geographical distribution.

The fishery for young and adolescent herring
In addition to the fishery for adults, a fishery for young and adolescent herring has taken place on the Norwegian coast and in the fjords, mainly in northern Norway. This fishery has been based on small herring ("småsild"), i.e. mainly 0- and 1-group fish, and fat herring ("feitsild"), i.e. II–IV group fish (Dragesund 1970). An analysis of the total landings shows that considerable fluctuations have occurred (Figure 4.4), starting at a level of about 60 000 tonnes during the first decade of the last century. The level then rose to 110 000–120 000 tonnes in the following twenty years, with a further increase during the 1930s to 180 000 tonnes. The landings then dropped slightly in the 1940s, followed by a significant increase from 1950 to 1959, when catches amounted to nearly 250 000 tonnes. The high level of exploitation continued during the 1960s, and reached a record of some 550 000 tonnes in 1967 and 400 000 tonnes in 1968, but thereafter the landings dropped rapidly to less than 50 000 tonnes in 1969 and less than 15 000 tonnes in 1971. Since 1974, catches of young and adolescent herring of this stock have been insignificant.

Before 1910, when the development of the fish meal industry started, most catches of young and adolescent fish consisted of fat herring. At that time, most of the small herring were used for canning, and were landed in southern Norway where most of the canning companies were located. After the start of the industrial herring fishery, a gradual increase in the landings of small herring took place, from a level of 20 000 tonnes in 1900–1910 to some 250 000 tonnes in 1960–61. The fat herring fishery did not show a similar increase in catches as the small herring. Landings from this fishery rose from 40 thousand tonnes in 1900–1910 to about 70 000 tonnes in 1950–1959, with periods of substantially higher yields in these fisheries in the 1930s and 40s. However, during the 1960s, a significant increase took place, and this fishery exceeded the small herring fishery with more than 400 000 tonnes in 1968 (Figure 4.4). In the beginning of the last century, most of the catches were taken by beach seine, when the herring penetrated into the fjords. However, gradually more herring were caught by purse seine, and from 1964 onwards this technique dominated in the fat herring fishery. The large quantities taken in 1968 were mainly from offshore waters in the Barents Sea.

The most important small-herring fisheries occurred in the fjords of northern Norway from late autumn to early spring. The fishery started with the immigration of 0-group herring to the fjords and continued throughout the wintering period. A peak in the early spring coincided with the emigration of 1-group herring from the fjords. The adolescent herring (fat herring) were usually found offshore along the Norwegian coast. However, in some years, during the late summer and early autumn, the herring migrated into the fjords and gave good catches. Historically, some fjords in northern Norway are well known for their rich fat herring fishery, e.g. Eidfjord in Vesterålen and Vestfjorden in Lofoten.

When the purse seine technique was introduced in the fat and small herring fishery in the 1960s, the fleet extended its operations to more offshore waters, mainly off Finnmark and Vesterålen/Lofoten. In these areas, the fishery started in early summer and reached a peak during late summer and early autumn. Further south along the coast (off Møre and Trøndelag), most of the fat herring were landed in spring and early autumn. The sharp increase in exploitation of young and adolescent herring during the 1960s, seriously reduced recruitment to the stock.

Concluding remarks
In this first period of the 20th century, the Norwegian spring-spawning herring fishery was characterized by the use of passive gears, such as drift nets and standing nets. Purse seining developed and grew in importance, but the usefulness of the method was limited because the fishermen were dependent on fair weather and in-shore conditions. The exploitation of the stock increased slowly through the first half of the century, and is believed to have had a rather small impact on the dynamics of the stock prior to 1950. During the 1950s and 60s, the fishery developed significantly in terms of both gear and fishing techniques that made the fishermen much more effective. From the mid 1960s, Norwegian spring-spawning herring were fished all over their area of distribution. The exploitation increased drastically and had a severe impact on the status of the stock. Its collapse led to a change in perception at the Institute of Marine Research regarding how rapidly a fishery could affect the status and dynamics of a fish stock.

4.4 Distribution and migration
Our understanding of distribution and abundance of herring in the northeast Atlantic region has developed greatly through the past 100 years. It is not possible to deal fully with the detailed history of these studies. However, the results of the investigations clearly show that there are persistent differences between some of the biological characters, such as the morphology and ecology of reasonably well defined groups of herring living in the northeast Atlantic. In the 19th century, the stock concept was not known, and neither the fishermen nor the scientists who were dealing with marine biology, possessed scientific knowledge of the fluctuation in fish stocks, or of whether the variability in availability of the fish could be ascribed to differences in stock levels or to changes in geographical distribution.

For centuries, however, the Atlanto-Scandian herring were the basis for herring fisheries, and in Norway scientific herring studies have been carried out since 1857 (Boeck 1871). Boeck brought together many historical facts about the Norwegian herring fishery. He believed that the fishery was periodic and that herring occurred in periods of high abundance alternating with periods of extreme scarcity. By the end of the 1860s, the yield of "spring herring" had decreased seriously, and during the next three decades, only small catches were taken along the west coast of Norway. Boeck's opinion that the herring would gradually leave their usual spawning grounds at the end of a herring period, created great anxiety among fishermen.

Later, Sars (1879) succeeded in drawing a fairly correct picture of the life history of the Norwegian spring-spawning herring. He could not accept the general view held at the time that herring were stationary and after spawning lived in deeper water just outside the coast, close to the spawning grounds. He believed that after spawning, the "spring herring" lived in the surface layers of the open sea between Scotland, Norway and Iceland, feeding mainly on copepods, and attaining maturity at an age of about six years. The spawning area was thought to be off the Norwegian coast between Stavanger and Kristansund, from which the larvae were spread northwards by the currents. Instead of many small herring populations located along the coast, Sars believed there were only two: the "spring herring" in southern Norway and the "large herring" in northern Norway.

The "large herring" caught in northern Norway in 1868–1874 approached the coast at Vesterålen and Lofoten and penetrated farther into the fjords in the autumn (October), and could be taken by beach seine. Both Boeck and Sars described this herring to be of the same size as the "spring herring", but in much better condition. In December, the "large herring" were in pre-spawning condition. Sars (1879) originally regarded this herring as a special tribe, whose spawning grounds were not known. Later he concluded that the connection between the "spring herring" and "large herring" fisheries was closer than he had previously believed.

During fieldwork carried out by Sars (1879) in the summer of 1872, many 0-group herring were found in the "spring herring" area, but almost no "spring herring" were caught. The observations led Sars to believe that the herring had spawned in the area, farther offshore than usual. He thought that they would probably soon return to the old spawning grounds, and questioned whether herring periods really did exist in the Norwegian herring fisheries, finding it unlikely that there should be any connection between the "spring herring" and the herring responsible for the great herring fishery in Bohuslän, Sweden (Sars 1879).

Sars also offered an explanation for the poor catch of "spring herring" at the end of the 19th century. He pointed out that one of the historical facts Boeck had brought to light was that a "Norwegian" herring period ends when the herring arrive at the coast later each year, and that this might be a result of the distance the herring have to cover during their spawning migration. If the distance is great, the herring will arrive late, the season will be short, and the catch accordingly poor, but if the herring had only a short distance to cover, they will arrive early, the season will be long, and the availability of herring better.

Hjort performed surveys of Norwegian waters and the Norwegian Sea at the beginning of the 20th century with RV "Michael Sars", and formed ideas about the migrations of the herring. Even though he initially seems to have held the idea that the variation in availability of herring was due to migrations

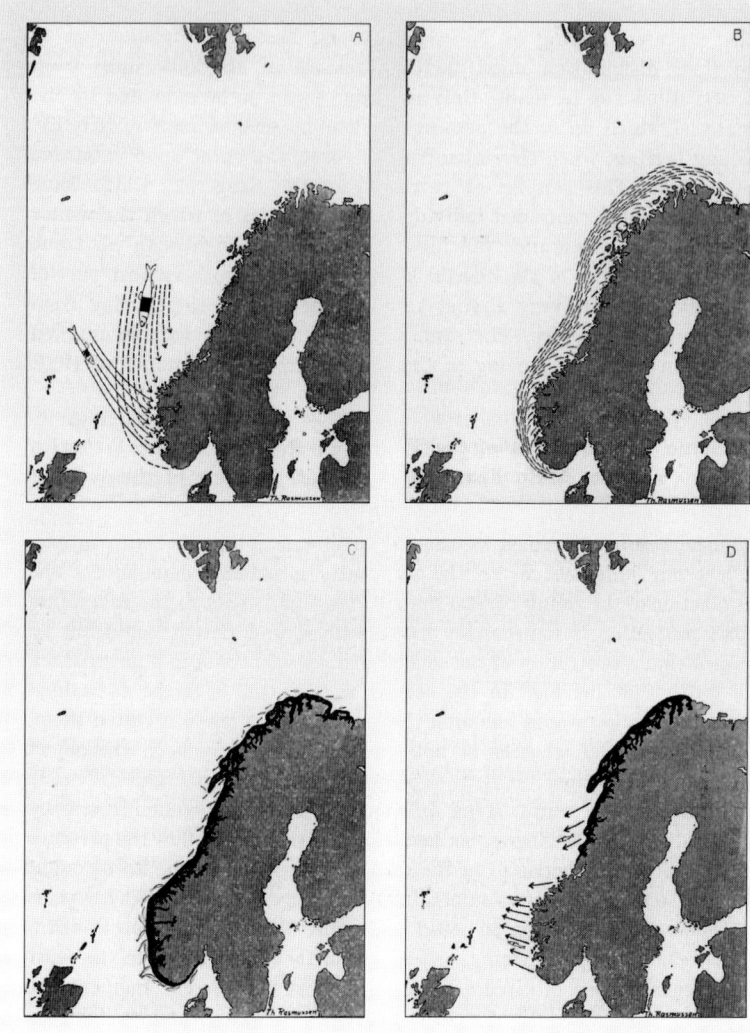

Figure 4.8
The migration cycle of Norwegian spring-spawning herring as described by Lea (1929b).

and not variations in abundance, he soon formulated ideas about variations in abundance of the different year-classes as reasons for differences in stock size (Hjort 1914). In 1907, Hjort and his colleagues established reading of herring and cod ages as a standard procedure to follow the strength of the different year-classes of the most important fish stocks. For Norwegian spring-spawning herring, scales were found to be most reliable for age determination, and after some years of studying herring scales and their characteristics, it was found that they could tell us a lot more than merely the age of the fish. They also provided information on growth, maturity, and where the herring had spent the immature part of its life, as described by Lea (1929a) in his paper on the herring scale as a certificate of origin.

Johansen (1919) found certain similarities between the herring off Iceland and those in Norwegian waters, and he suggested to call them Atlanto-Scandian spring herring, a term that later was changed to Atlanto-Scandian herring when the Icelandic summer-spawning herring was included.

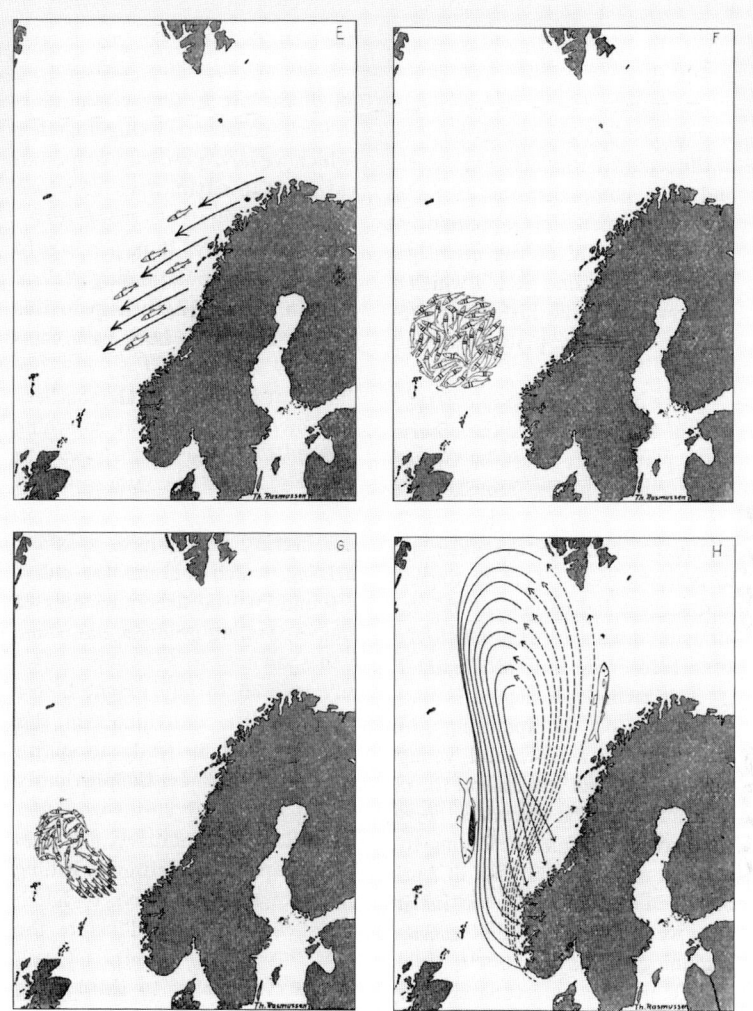

Figure 4.8 (cont.)

Understanding of the geographical distribution based on studies of scales

Through the first decades of the 20th century, techniques for ageing herring scales were developed (Broch 1906; Dahl 1907), and this helped the scientists, not only to follow cohorts of the herring stock, but also, by studying the characteristics of the annuli, enabled them to describe where the herring had spent its first years of life. Lea (1929a) performed detailed studies of the structure of herring scales, and found certain characteristics in the scale (size, colour and width of the annuli) that depended on where the herring had been during the different phases of its life-cycle. On the basis of these studies, Lea (1929b) produced detailed descriptions of the migration and distribution of the Norwegian spring-spawning herring, and formulated ideas of an oceanic stage based on scale structure.

Lea described the following migration cycle (Figure 4.8);

A. Arrival of the spawning herring at the spawning grounds in two main groups: first, the older spawners, and then the recruit-spawners, during the season January–April.

B. Dispersal of the year's fry along the entire coast.
C. Sojourn off the coast in the herring's second year of life.
D. Departure from the coast, and transition to the oceanic stage, in their third year of life, of the herring which have lived along the southern part of the western coast.
E. Departure from the coast and transition to the oceanic stage, mainly in their fourth year of life, of the herring which have lived further north along the coast.
F. Oceanic stage. Duration according to circumstances, from one to three years, age of herring between two and six, possibly seven years.
G. Abandonment of the oceanic stage, after sexual development, by the herring which will become sexually mature in the forthcoming spawning season.
H. Annual migration of the adult herring between its spawning grounds and its summer haunts.

Lea (1929b) found that the zones of the scales (the winter rings and the summer rings) had certain characteristics depending on where the herring were spawned and where they had spent their early life (on the coast or offshore). The basis for the system was that young herring in coastal areas of northern Norway have sharp winter rings (2–6 rings) and slow growth, and are classified as northern type (N). Herring that grow up on the south-west coast have more diffuse winter rings, 1–2 rings, seldom 3, and faster growth; these are classified as southern type (S). An important finding was the oceanic stage (Lea 1929b). When the young herring leave the coastal areas, they migrate westwards into the central part of the Norwegian Sea. Here they spend one or two, or seldom three years, before they mature and join the spawning stock. He demonstrated that the development from fat herring to adult herring took place through an intermediate oceanic stage before they appeared on the spawning grounds. Lea

Figure 4.9
Norwegian spring-spawning herring scale, described by Lea (1929a) as a certificate of origin.

(1929a) and Runnstrøm (1936) showed that the onset of maturation in Norwegian spring herring causes visible changes in the ring structure of the scale. From the late 1920s to 1970 not only the age of the fish, but also age at maturity, was recorded (Beverton et al. 2004).

After spawning, the herring moved northwards to feed in the Norwegian Sea, and in May–June were observed at the latitude of Lofoten and were sometimes caught near the coast. In July, they were observed off Bear Island, and in August–September, as far north as off Spitsbergen. In October, the herring were on their way south again on a level with Lofoten, and in the winter months further to the south. This description was based on observations made on the coast of Lofoten, and from material acquired during expeditions in the northeastern part of the Norwegian Sea, west of Bear Island and Spitsbergen. The herring scale analysis enabled Lea to confirm that these herring belong to the Norwegian spring-spawning stock.

The migratory pattern and the oceanic stage, indicated by Lea (1929b), seem to be in accordance with Russian investigations during the 1930s (Marty and Fedorov 1963). These observations showed that adult herring were distributed in the central and north-eastern parts of the Norwegian Sea during the feeding period.

Understanding of the geographical distribution based on tagging experiments
At the time at which Lea published his observations of the migration route of the adult herring, he was not aware of the feeding migration to the western part of the Norwegian Sea and to the area off north and north-western Iceland. Árni Fridriksson, in 1935, was of the opinion that the herring caught during the summer off the north coast of Iceland were at least partly identical to the Norwegian winter herring. He found it desirable to carry out tagging experiments in order to prove or disprove the theory of long-distance migration (Fridriksson 1944). A joint project involving Icelandic and Norwegian fishery scientists was organized just after the World War II, and between 1948 and 1950, a total of 42 054 herring were tagged (Fridriksson and Aasen 1950, 1952). The experiments proved that, at that time, the Norwegian spring spawners migrated to Iceland during the summer and returned to the Norwegian coast to spawn during the winter. However, the migration routes between these two areas were not known in detail.

Joint surveys in the Norwegian Sea
The 1948 Statutory Meeting of the ICES recommended that countries interested in the exploitation of this stock should adhere to a common plan of investigations to be coordinated by the Chairman of the Northwest Area Committee. In adhering to this recommendation, Denmark, Iceland, Norway, Scotland and Sweden carried out research in the area in 1949 (Jakobsson and Østvedt 1999). The recommendation from 1948 can be regarded as the first ICES initiative for joint international herring surveys. At the 1951 ICES meeting in Amsterdam, Denmark, Iceland and Norway agreed to cooperate in an extended herring research programme to cover the Norwegian and Iceland Seas. Beginning in 1957, the USSR also participated in this programme. The surveys, which were carried out during the summer, took place annually until 1970. Joint meetings of scientists participating in the surveys were held at the end of June every year (Jakobsson and Østvedt 1999). These investigations gave a fairly good

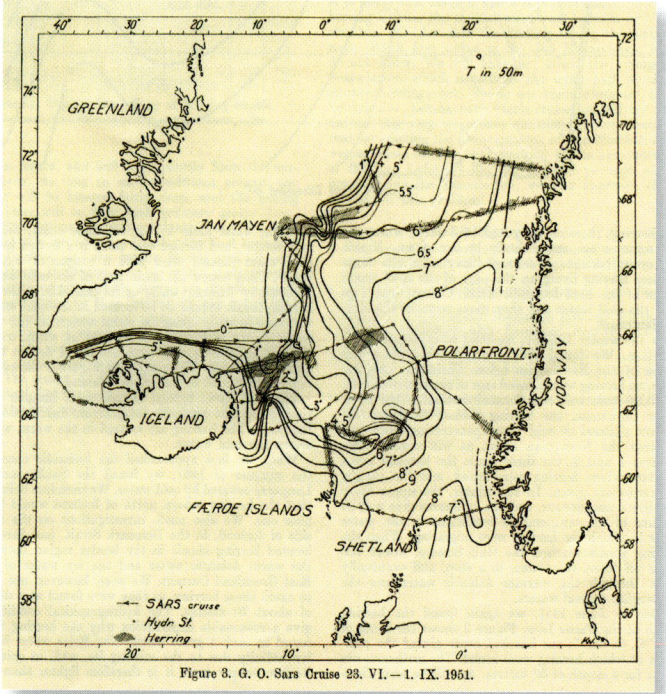

Figure 4.10
The geographical distribution of Norwegian spring-spawning herring as mapped by Devold (1952) in the Norwegian Sea, 1950.

Figure 4.11
Descriptions of the migration pattern of herring in the Norwegian Sea (Devold 1963).

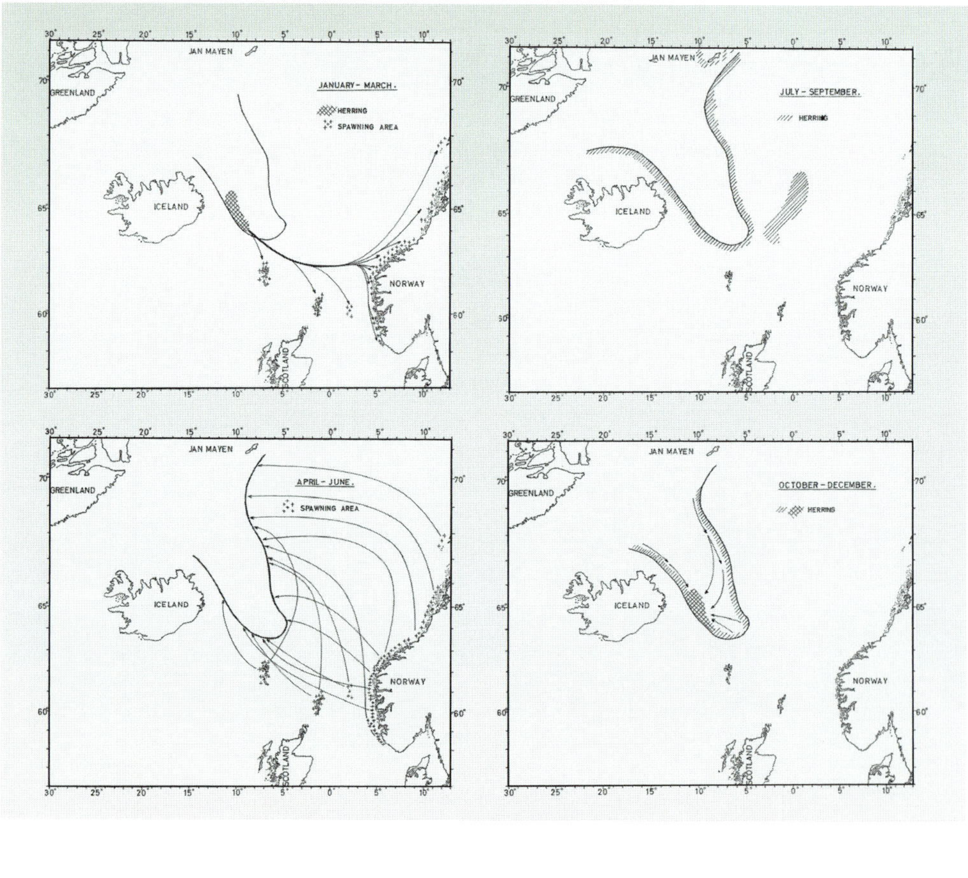

indication of the summer distribution of Atlanto-Scandian herring in relation to temperature and plankton distribution (Østvedt 1965).

Understanding of the geographical distribution based on acoustic surveys
During the 1950s and 60s, the Norwegian herring scientist Finn Devold carried out annual acoustic surveys in the Norwegian Sea during summer and winter, and he showed that small herring of the rich 1950 year-class were distributed far offshore in the Norwegian Sea (Figure 4.10). He suggested that only a part of the total 0-group population entered the Norwegian fjords. The RV "G.O. Sars", led by Devold, mapped the distribution of the herring as they migrated from their over-wintering areas east of Iceland towards the Norwegian coast to the spawning grounds. The Norwegian herring scientists at the time had a very good relationship with the fishermen, as Devold appeared to be herding the herring towards the Norwegian coast. Over the radio, the scientists guided the fishermen to the richest concentrations of herring. Based on the results of these surveys, Devold (1953) produced maps of herring distribution in the Norwegian Sea throughout a whole year (Figure 4.11).

In 1959 and during the 1960s, Dragesund (1970) carried out studies of the distribution and abundance of young and adolescent herring on the Norwegian coast and in the Barents Sea. Dragesund described where the young herring were distributed at various stages. He found that in the autumn, the 0-group of Norwegian spring-spawning herring was always to be found in the fjords of

Finn Devold

Finn Devold (1902–1977), the leading herring scientist in Norway through the 1950s and 60s, had a special theory regarding the distribution of the Norwegian spring-spawning herring and how the distribution changed throughout the herring periods (Devold 1963). He was strongly of the opinion that in the course of a herring period, they gradually spawn further and further north along the Norwegian coast. The herring also occur on the coast gradually later in the winter season and start spawning later as the herring period develops. According to Devold, by the end of a herring period, small numbers of herring spawn off the coast of Northern Norway. By that time, the stock is in a phase of low abundance and consists of a mixture of small immature and large herring. After a year or two of very low spawning activity, the herring, according to the theory, start to migrate to the Bohuslän coast and spawn off the west coast of Sweden. Devold tried to justify the theory through a temperature-related shift in maturation, which, year by year, caused the herring to migrate further south in order to find suitable spawning temperatures. This hypothesis was strongly contested by a number of scientists, among them Høglund, in Sweden, who had sampled and analysed the remains of herring from the Bohuslän coast from previous production. He found that these remains originated from North Sea herring and therefore did not believe in Devold's theories (Høglund 1959).

In retrospect, the Bohuslän herring have been verified as being of North Sea origin. Corten (1999) proposed a mechanism for the Bohuslän herring periods, which fits well with the understanding that the North Sea herring overwintered at the Bohuslän coast when the Norwegian spring-spawning herring were low in abundance (between the herring periods). However, Devold was probably right in his observations of the Norwegian spring spawners migrating further north as the herring period progressed.

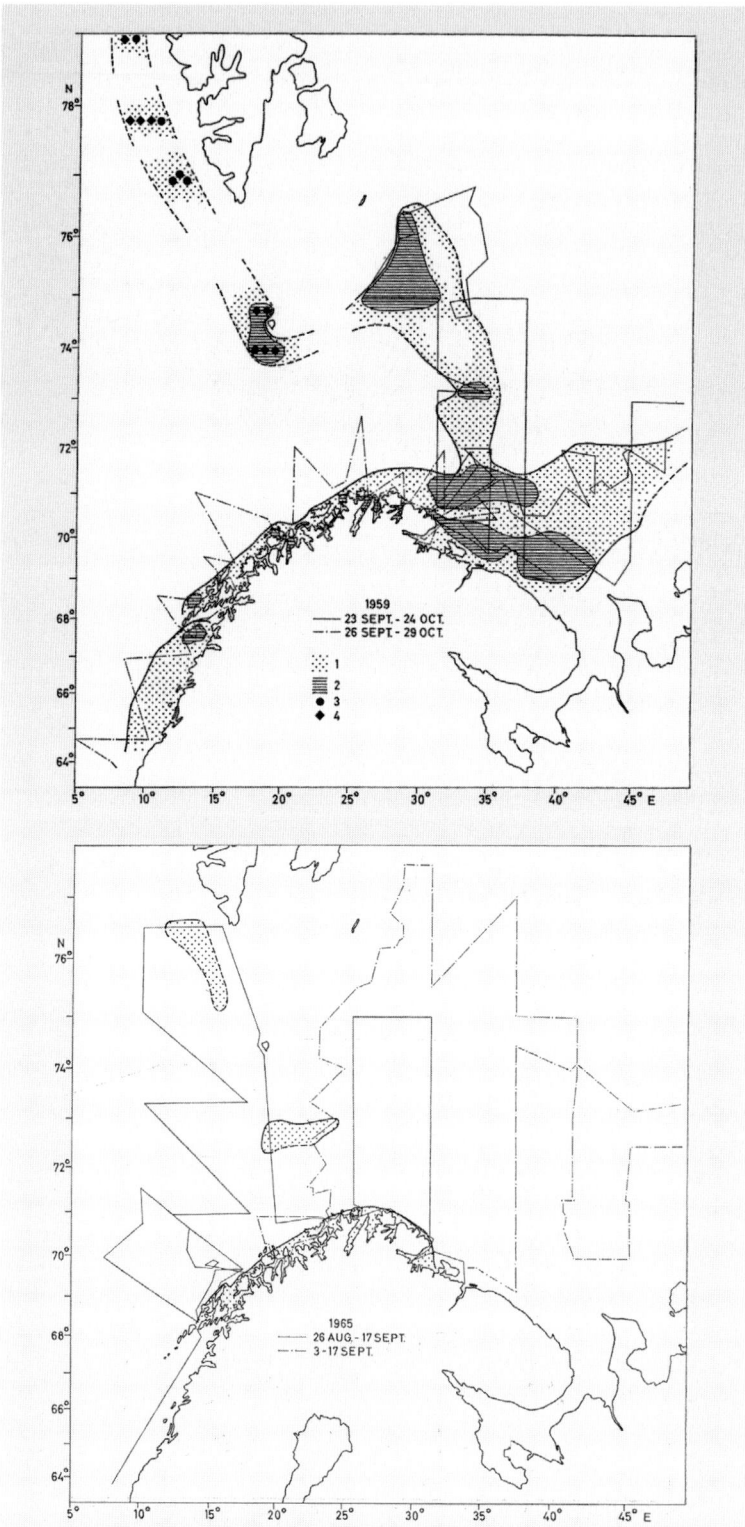

Figure 4.12
The geographical distribution of 0-group herring for year-classes of different strength (Dragesund 1970b).

western Norway and off northern Norway. When rich year-classes occurred, they were also distributed offshore, mainly in the Barents Sea, but also in the area west of Bear Island and Spitsbergen. The 0-group herring entering the fjords during the autumn, remained in the fjords the following winter and spring, and then gradually emigrated and mixed with young herring offshore. The young and adolescent herring, which were distributed offshore and in the Barents Sea, remained there for one to four years, depending on their rate of growth, and then migrated westwards, either south along the Norwegian coast or further westwards into the Norwegian Sea (Dragesund 1970b) to join the adult stock. Dragesund carried out tagging experiments of adolescent herring off the northwest coast of northern Norway, suggesting this migration route. Later studies confirmed this distribution cycle. The young and adolescent herring are spread, ranging from the fjords of northern Norway to the open ocean of the Norwegian Sea and the Barents Sea, and their distribution varies with the strength of the different year-classes. In 1970, Dragesund described the geographical distribution of 0-group herring for year-classes of different sizes. Examples are shown in Figure 4.12.

4.5 Stock composition and growth
Stock units

The name "Atlanto-Scandian spring herring" was introduced by Johansen (1919) to describe most of the spring-spawning herring found along the Norwegian coast, on the northeastern edge of the North Sea, in the Norwegian Sea off Iceland and off the Faroes. The methods for stock separation were considerably improved by the use of scales and growth. The use of herring scales for growth and stock structure studies and as a "certificate of origin", was fully described by Lea (1929). The studies of stock structure in the 1920s and 30s carried out by Lea and by Runnstrøm, were fundamental in terms of establishing our knowledge of the biology of Norwegian spring-spawning herring. However, a full understanding of the stock composition of the Atlanto-Scandian herring was not in place before the results of the tagging experiments in the early 1950s made the division between the Norwegian spring-spawners and the Icelandic summer- and winter-spawners clear.

Runnstrøm (1936, 1941) continued Lea's studies, and further described the structure and growth of the herring stock, depending on the migration and geographical distribution of the various age groups.

The two Icelandic herring stocks, the summer-spawners and spring-spawners, belong to the Atlanto-Scandian herring group together with the Norwegian spring-spawners (Johansen 1919). Detailed investigations have shown that the Icelandic summer-spawners differ considerably from the spring-spawners in a number of morphological and physiological characters (Johansen 1926; Fridriksson 1944, 1958; Liamin 1959; Einarsson 1951). Smaller differences have been observed in the characters between the Norwegian and Icelandic spring-spawners, and the Icelandic spring-spawners were difficult to distinguish from the southern growth type of Norwegian spring-spawning herring (Jakobsson *et al.* 1996). On the basis of growth and age at first maturation, Østvedt (1958) concluded that it is evident that the Norwegian spring-spawning herring are homogeneous as regards racial characters. He further concluded that it is evident that the northern and the southern growth types cannot be defined as different races, and that the northern type dominates rich year-classes. Seliverstova (1970) related the presence of northern and southern growth types to the strength of the 1950 and 59 year-classes, and found that the northern growth type dominated in the more abundant 1950 year-class. She also related maturation of certain age groups in the spawning stock in 1959 to hydrological conditions.

Extensive tagging experiments, carried out on a national basis together with the acoustic investigations carried out both in Norwegian and Icelandic waters, and studies of the genetics in various herring stocks in the Northeast Atlantic, have confirmed the structure of the three Atlanto-Scandian herring stocks.

The general idea of herring biology through the first decades of the last century, was that herring which were distributed in different areas was of the same origin. The stock concept was not developed, and growth and differences in geographical distribution were not well understood. However, through the studies of scales and the development of research on growth through studies of scales, the understanding of different herring stocks and components within stocks developed. The sampling of the Norwegian spring-spawning herring, which was initiated in the early years of the 20th century, and the studies of scales and otoliths by Broch and Dahl (Broch 1906; Dahl 1907), soon gave results. Hjort and Lea (1911) discovered the new world of demography applied to fishery biology, and made studies on length frequencies and growth types of different age groups along the Norwegian coast. In "The Fluctuations in the Great Fisheries of Northern Europe", Hjort (1914) delivered the following message: "By large-scale stock age compositions one can identify stocks, classify them by age, measure their growth, and make predictions of changes in their fishable biomass". Documentation from these early years confirms that scientists were already aware of central structural features of the Norwegian spring-spawning herring. However, because of the disagreement between British and Norwegian scientists regarding the use of scales and otoliths to determine age, it was not until 1923 that the question was settled and the usefulness of scales and otoliths was accepted in the ICES community (Chapter 2), and used in studies of North Sea herring.

Growth and maturation
Growth and recruitment are important fields of research for fisheries science, leading directly into fish stock dynamics and advice on fisheries management and understanding of the general biology. The studies by Lea (1929) on the growth of herring studied by the patterns of the zones of the fish scales have already been mentioned. Runnstrøm (1936) continued these studies and described the northern and southern growth types of the Norwegian spring-spawning herring.

At the ICES Statutory Meeting in 1964, Østvedt (1964) presented a paper on the long-term changes in growth and maturation of Norwegian spring-spawning herring. The paper analyses the long-term changes of the mean length for different age groups. One of its conclusions is that the mean length at age has tended to increase through the period 1908–1963. The increase in mean length can be traced from the youngest age groups in different growth types of herring (N or S). Increases in growth were also observed in the North Sea herring stocks (Cushing and Burd (1957), and could have been caused by common environmental factors. It has been shown that there was a rise in temperature in the Northeast

Atlantic through this period (Toresen and Østvedt 2000). The general increase in exploitation in the course of the century may also have affected growth rates. It has also been shown that growth in the Norwegian spring-spawning herring stock is density-dependent, and that maturation is dependent on growth (Toresen 1990; Toresen and Østvedt 2002), and as exploitation has increased, the density of the stock has decreased. During the summer season, growth is dependent on the temperature conditions in the Norwegian Sea (Holst 1996).

Recruitment

Recruitment studies were started long before stock management was introduced in the herring fisheries. Recruitment studies were carried out in order to obtain an idea of the incoming year-classes and how the fishery would develop in the coming years. A basic question in the study of the Norwegian spring-spawning herring has been which factors influence the year-class strength. This stock has large variations in year-class strength, and several scientists have been occupied with the question (Hjort 1914; Dragesund 1970a; Dragesund and Nakken 1973; Toresen and Østvedt 2000). Dragesund (1970a) considers the following factors to be most important in determining year-class strength: 1) the extent of the spawning area, 2) the duration of the main spawning season, 3) the rate of dispersal of larvae from the spawning grounds, and 4) the coincidence in time between the availability of suitable food and hatching of herring larvae. Dragesund and Nakken (1973) concluded that the results may indicate a relationship between the parent stock size and subsequent abundance of the resulting year-class when favourable conditions for spawning and hatching exist. However, they point out that in most years, year-class strength was determined by factors other than the size of the spawning stock. Toresen and Østvedt (2000) found that recruitment was related to temperature conditions in the inflowing Atlantic water through the Kola section (Figure 4.13), and that at low temperatures, the probability of good recruitment is very low, while at higher temperatures, both rich and poor year-classes occur. The underlying mechanism in the correlation between temperature and recruitment is as yet unclear. There may be both direct and indirect effects. Higher temperature has been found to positively affect growth in fish larvae (Fiksen and Folkvord 1999). Higher temperature has also been found to positively affect the production of zooplankton populations on which

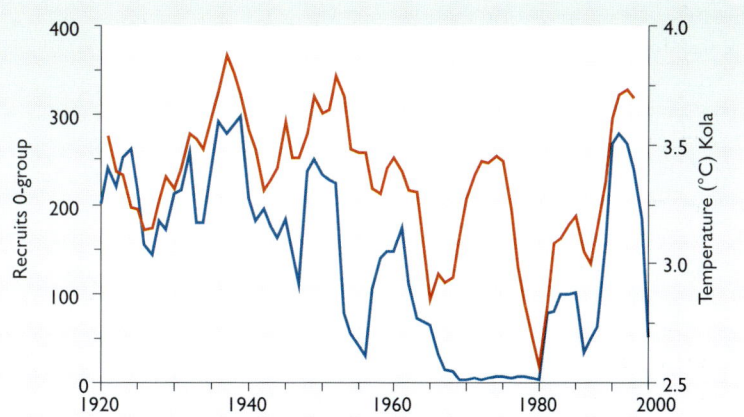

Figure 4.13 Temperature (red) and recruitment of Norwegian spring-spawning herring (blue) (Toresen and Østvedt 2000).

the herring feed (Aksnes and Blindheim 1996). However, the pattern of the relationship between recruitment and temperature (poor recruitment at low temperatures and both high and poor recruitment at high temperatures) indicates that the indirect effects are stronger, since high temperature is necessary but not sufficient for good recruitment (Toresen and Østvedt 2000). This feature has also been found in Northeast Arctic cod (Sundby 1994).

4.6 Exploitation, advice and management

In the first half of the last century, scientists dealing with fish and fish stocks were mainly occupied with biological research and studies related to the distribution and migration of the fish. The stock concept was poorly developed, and so was the idea that the fisheries could make any impact on the abundance of fish. In addition, by 1950, fish abundance estimation and mapping were only at a very early stage of development, nor was analytical fish stock assessment highly developed. Therefore, knowledge-based advice from the scientific community for Norwegian spring-spawning herring was not yet sufficient to allow such management of the fisheries, as could reduce the exploitation of this stock if that should be necessary.

Fisheries management as we know it today, with a quota regulation system based on annual assessments and allowable catches, has come into force in the course of the past 30 years. Before about 1970, fisheries management was nearly absent in the Norwegian fisheries. During the 1960s, assessment of fish stocks was still an emerging science, and regular assessments of fish stocks started to come into effect during the late 1960s and the 1970s.

The first important milestone for the development of fish stock assessment was the understanding of year-classes as units that vary individually in abundance (Hjort 1914). This understanding was based on the discovery of the possibility of determining the age of fish by otoliths or scales (Smitt 1895; Hoffbauer 1898), and Smitt was probably the first to notice a connection between the scales of herring and their age (Dahl 1907). Heincke informed Hjort of these findings, whereafter Broch and Hjort visited Heincke's institute in Helgoland, Germany (Schwach 2000). The use of age determination as a standard technique in fish sampling started with the reading of herring scales, and regular sampling of herring scales started at the Institute of Marine Research (IMR) in Bergen in 1907. Broch was one of the pioneers of herring scale studies, and was the first to realise that the scales contained information on the annual growth of the fish and that there were systematic differences in the zones of the scales from different races of herring. When Hjort studied Broch's report on his findings in 1906 he said: "Oh damn it, Broch! You probably don't understand the scope of what you just have found out" (Schwach 2000). Hjort was probably right. These findings opened up demographic thinking in the assessment of fish stocks, and have been a central approach to any fish stock assessment ever since.

However, as early as in 1924, Lea published a note on the exploitation of "small herring". Fishermen harvesting the somewhat older "fat herring" were worried about catches of small herring and asked whether the "small herring" fishery was influencing the yield of "fat herring" in subsequent years. Lea concluded that it was unlikely that catches in the fat herring fishery would rise if the small herring fishery were regulated. Although both Lea (1924) and later Devold (1963) did not find any connection between catches of small herring and subsequent yields of the fat herring fishery, and thus could not recommend regulation of the 0-group herring fishery, many Norwegian fishermen have maintained that the exploitation of small herring affected both the fat herring fishery and the fishery of adult herring.

In 1929, ICES organized a special meeting on "Fluctuations in the abundance of the various year-classes of food fishes" (ICES 1930). In relation to herring research, one contribution was quite innovative and demonstrated one of the most important future uses of age determination: Lea's "Mortality in the tribe of Norwegian herring" (Lea 1930). This was the second important milestone towards analytical assessments for the Norwegian spring-spawning herring. He based his estimates on the presence of the different year-classes, year after year, in samples from the stock. He found that the presence of given year-classes in the samples from the spawning grounds decreased at a certain rate. He found that this rate (total mortality rate of 0.21) varied only a little between 1907 and 1926 (Figure 4.14). The exploitation of the stock was at a low level during this period (Toresen and Østvedt 2000), and the assumption that natural mortality is low (less than 0.2) for the adult part of the stock, and is also relatively stable, seems to be valid.

The serious declines in some of the most important European herring fisheries during the 1950s, especially those of the Norwegian west coast and in the North Sea, caused fisheries scientists to pay attention to the potential importance of man-made factors as major determinants. ICES, acting on a recommendation from its Herring Committee (1960), therefore held a Herring Symposium in 1961 in order to bring out all the available knowledge about the subject. The

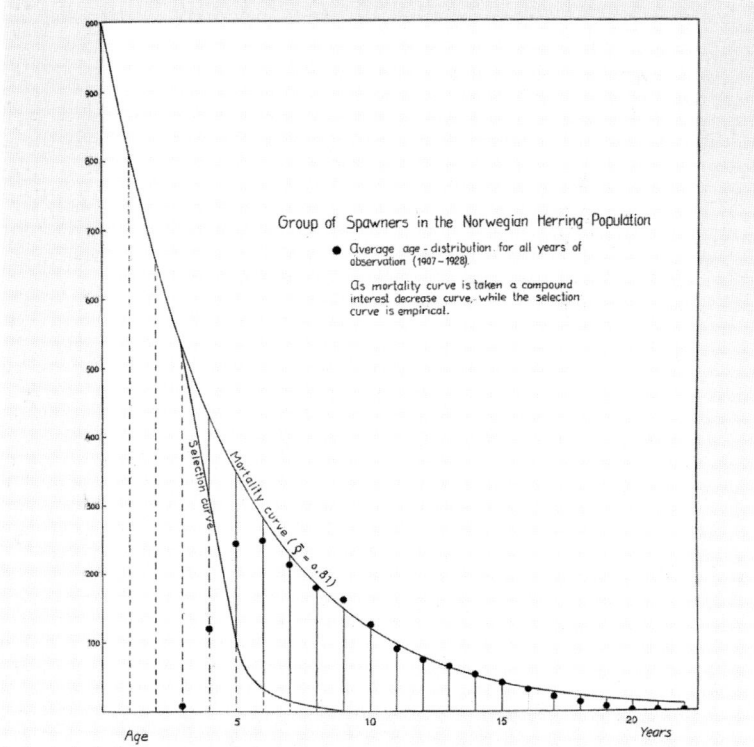

Figure 4.14
Mortality of year-classes of Norwegian spring-spawning herring (Lea 1930).

meeting did agree that recruitment had failed, but there was disagreement regarding the causes of that failure. Marty and Federov (1963) held the view that the juvenile fishery should be reduced in order to improve recruitment. This was heavily debated and strongly opposed by Norwegian scientists. Another paper presented at the symposium, on the state of the Atlanto-Scandian herring (Dragesund and Jakobsson 1963), was based on herring tagging returns since 1948, and was a joint project by Norway and Iceland. Their estimate of the stock of about 20 million tonnes of adult herring in 1952 was high, probably because it was based on the very early tagging experiments, in which tagging mortality may have been much higher than in later years when more experience had been gained in handling the herring. Comparing these stock estimates from 1953 to 1959 with the results of the most recent population analysis (ICES 2001), it is clear that the results from the tagging experiments at the time reflected well the decline in the stock during 1953–1959, and CPUE data indicated the same (Østvedt 1963). The tagging results overestimated the stock size during the three first years, but in 1956, the estimates were equal, while the tagging results gave somewhat lower stock size estimates and a sharper decline in 1957–1959. However, the results of the tagging experiments do not necessarily reflect the changes in abundance of the spawning stock as well as the stock analysis done by the ICES working group where additional information is used. The results of the tagging may have represented only the available stock of herring in the adult fisheries off the north coast of Iceland and the west coast of Norway during the winter season. Comparison of the abundance of five-year-old and older herring estimated from the stock analysis in the ICES working group with that calculated from the tagging experiments, shows an even closer agreement than comparison with the spawning stock. Thus, as early as 1961, the decline in the stock during the 1950s was clearly illustrated in the paper presented by Dragesund and Jakobsson (1963) (Figure 4.15).

In accordance with the recommendations of the Herring Symposium in 1961 and endorsed by the 1961 and 1962 ICES annual meetings, it was decided to establish an assessment working group for Atlanto-Scandian herring. The first meeting of the group was in 1963, with the following terms of reference:

"In view of the recent serious decline in the fisheries based on the Norwegian spring-spawning tribe and of the complexity of the scientific problems involved, the Symposium recommends that a Working Group be set up, composed of

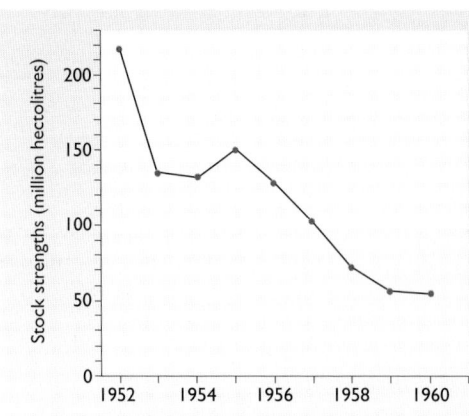

Figure 4.15
The decline of the stock of Norwegian spring-spawning herring as described by Dragesund and Jakobsson in 1961 (Dragesund and Jakobsson 1963).

representatives from all countries with fisheries based on this tribe, to examine all available data relating to the elucidation of its causes, including the part played by the fisheries for juvenile, pre-recruit herring" (ICES 1963).

These terms of references clearly indicate that early in the 1960s, there was a sense among the scientific community that the exploitation of juvenile and pre-recruit herring could be important for the decline in the stock.

Although the early ICES working groups dealing with the assessment of the Atlanto-Scandian herring (ICES 1963, 1964, 1965a) were preoccupied with biology, estimates of stock size and how the stock decreased in abundance throughout the 1950s were presented. Comparisons of the results of the tagging and acoustic data with those of the stock analysis in the ICES working group up to and including 1964, show that the estimates presented by the early working groups on stock abundance were very close to those obtained by the most recent results of the analysis for five-year-old and older herring. Similarly, the working groups presented fishing mortalities for the 1950s and the early 1960s (ICES 1964) that were almost identical to those obtained from the recent analyses for 1953–1958. It is thus clear that the early ICES Atlanto-Scandian Working Group that met in 1963–1965 estimated the stock and the rate of exploitation realistically, although the fishing mortality appears to have been temporarily overestimated.

The severe reduction in stock abundance observed during the late 1950s and the early 1960s was correctly regarded as being due to a series of weak year-classes (1951–1958) recruiting to the adult stock. Whether this was due to the juvenile fishery was much debated at the time. Norwegian scientists disagreed among themselves on this point. However, the disagreement was not fully reflected in the Working Group reports. The 1964 Working Group report concluded that:

"while the recent fishery and stock decline could definitely be attributed to a sustained failure in recruitment, it was not possible from current evidence to determine the part played by the fishery for 'small' and 'fat' herring as a governing factor in it. However, the Soviet representatives expressed the view that this fishery has influenced the level of recruitment in recent years."

In the report of the *ad hoc* Assessment Group on Herring and Herring Fisheries in the North-Eastern Atlantic (ICES 1965b), the scientists concluded that the primary cause of failure of year-classes at the time to provide good recruitment was natural. It was probably on the basis of this report that the ICES Liaison Committee concluded that (ICES 1966):

"The exploitation of the adult Norwegian spring-spawners is probably still at a level where no benefit for total landings can be expected by any regulatory measure. The decrease in catch from 1957 to 1963 was caused by the sequence of poor year-classes from 1951 to 1958. The primary cause of the failure of any of these year-classes to provide good recruitment has been natural."

An interesting question is whether the state of the art of fish stock assessment was good enough in the 1960s to have been capable of preventing the collapse of the Norwegian spring-spawning herring stock. No analytical assessment (VPA) had been made for the stock, but it had been evaluated in ICES working groups. Sætersdal (1980) reviewed the stock collapse and wrote:

"Historically, the collapse of the Atlanto-Scandian herring stock was the first serious one in the northeast Atlantic and it also represents by far the largest loss of fishable biomass in the area. However, until 1970 the stock was dealt with only once by the Liaison Committee. A large stock decline was recorded in a 1965 assessment, but apparently scientists firmly held the view that this was all due to natural stock variations, and they advised that exploitation was still at a level where no benefit for total landings could be expected from any regulatory measure."

Sætersdal concluded that the scientists had poorly recorded the stock events, and that the responsibility for the lack of any attempts at management of the fishery during the period of the stock collapse lay entirely with the scientists. However, fisheries research in terms of monitoring the stock abundance and assessing the annual changes in abundance, did not develop as fast as developments in fishing technology, and therefore, the collapse of the stock came as a surprise for both fishermen, managers and scientists.

Even though some scientists were sceptical about the intensification of fishing throughout the 1960s, Sætersdal's conclusion was probably right. The state of the art at the time, in terms of assessment methods and stock diagnosis, was not ready for active fisheries management involving restrictions on the fisheries.

The development of fish stock diagnostics was at too early a stage for firm and concurrent advice on management in the present situation.

4.7 The period 1970–2000: management and stock recovery

The collapse of the Norwegian spring-spawning herring stock influenced research, advice and management. In retrospect, it has been concluded that the decline of the stock through the 1950s could not have been avoided, and the severe reduction in abundance at the time was mainly due to reduced recruitment. However, the collapse of the stock, which was mainly caused by the escalation of exploitation in the 1960s, could have been avoided if proper management measures had been implemented. However, it is important to bear in mind that the rise in exploitation was a result of the revolution in fishing techniques and gear technology, and the effect of these developments was difficult to foresee.

The period 1970–2000 represented a complete change in the fishery and management of the stock. Both abundance and availability were drastically reduced, and its geographical distribution was significantly changed compared to what was known about the distribution before the stock collapsed. In the early 1970s, herring were virtually absent, and only very small quantities were observed. The spawning grounds were at a few specific grounds off Møre, and the herring no longer migrated offshore and into the Norwegian Sea for feeding during the summer months. The herring stayed near the coast during the summer and autumn and wintered in a few fjords, off Møre and northern Norway. This all-year coastal distribution lasted throughout the 1970s and until the 1983 year-class migrated out of the Barents Sea as three-year-olds. In the late 1980s, the maturing parts of this year-class resumed feeding migrations into the Norwegian Sea after spawning. Since then, the mature herring have made feeding migrations in the Norwegian Sea during summer. However, these migrations have not been as far west as before the stock collapsed, and for a long period, in the autumn, the herring migrated back to the Norwegian coast where they wintered in fjords in northern Norway.

From 1970 on, there was a virtually complete halt in the fishery for about 15 years, and the fishermen, scientists and fishery managers faced new challenges to manage the fishery in a way for the spawning stock to rebuild itself as rapidly as possible. The fishing industry wished to reopen the fishery at an earlier stage, mainly because the herring were observed at the coast all year around. There was a lack of herring on the market because the North Sea herring stock had also collapsed in the late 1970s. It was therefore a difficult task for scientists and managers to convince the fishing industry that in order to rebuild the stock, fishing would have to be kept at a very low level, preferably with a fishing mortality of less than 0.05. There was a deep concern among the herring scientists that the fishery had been too intensive in the early 1970s. The role of the scientists had completely changed, from being guides to the highest concentrations of fish before the stock collapsed, to being conservers and advising the protection of the few herring left in the rebuilding period.

Main fishing fleets and exploitation
Throughout the 1970s and the early 1980s, there was very low exploitation of Norwegian spring-spawning herring. In 1973 and 1974, less than 10 000 tonnes were landed, and in the whole period 1975–1983, only small amounts of herring were landed by drift net and small purse seiners. The herring went to human consumption and bait. In the mid-1980s, when the abundance of the strong 1983 year-class was verified, a herring fishery on the Norwegian coast was reopened. Since then, there have been two main fisheries for herring: *a)* one that primarily exploited the mature adult herring during winter and spring, mainly by purse seiners, and *b)* a fishery for the same part of the population as fishery *a*, but carried out in the Norwegian Sea while the herring occupy their summer feeding areas.

In 1984–2000, the fishery expanded, and large purse seiners were allowed to exploit the stock. The fishery now takes place all year round, with a concentration of fishing on the spawning grounds and feeding grounds (Norwegian Sea) during the summer. Some smaller purse seiners and trawlers exploit the stock in the overwintering area. The drift net fishery ended in the late 1970s.

During the rebuilding phase, herring fisheries were kept at the lowest possible level and then allowed to slowly develop; purse seiners are currently the most important component of the herring fleet. Some trawlers from European countries and Russia harvest the stock in the Norwegian Sea during the summer.

During the late 1970s, the quantities landed were very low. Landings increased somewhat after 1985, and the mean fishing mortality rate (ages 5–12) increased to 0.8 in 1983, 0.4 in 1985, and 1.0 in 1986. Figure 4.16 shows landings and fishing mortality of Norwegian spring-spawning herring in 1970–2003. Landings subsequently fell to less than 100 000 tonnes in 1990 and 1991, while spawning stock biomass increased and fishing mortality decreased to low levels again by 1991. Throughout the 1990s, landings rose from 480 000 tonnes in 1994 to more than 1.2 million tonnes in 1996–1998. Since 1991, fishing mortality has been low and less than 0.15 (ICES 2004).

Landings of small and fat herring were insignificant after 1974. However, despite decreased catches after 1968, fishing mortality on the juvenile component remained high until 1973 (Toresen and Østvedt 2000). The exploitation of young

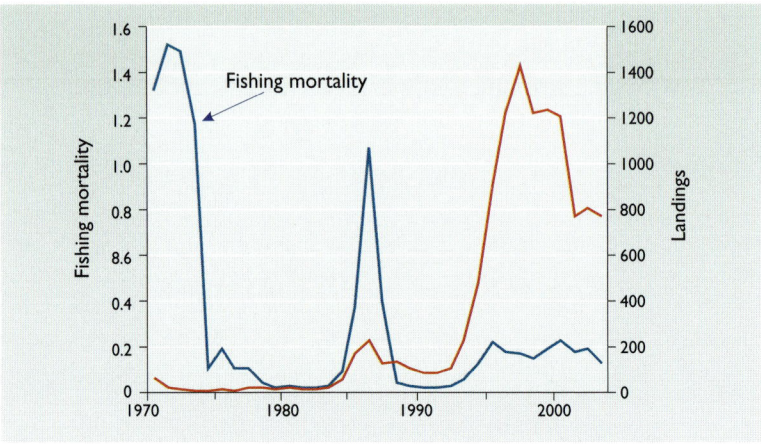

Figur 4.16
Landings and fishing mortality of Norwegian spring-spawning herring since 1970 (ICES 2004).

herring has been prohibited since 1977 due to a minimum size regulation of 25 cm, which has been strictly enforced.

Distribution and migration; changes compared to the pattern prior to the collapse
Our current understanding of the geographical distribution of the herring stock is a verification of what was believed to be the general distribution of the stock early in the last century (Lea 1929). Based on acoustic surveys in the different areas through the year, it has been confirmed that there are spawning sites along the west coast of Norway, but the importance of the various grounds varies and is dependent on the abundance of the stock and probably also on where the herring spend the winter. It has been confirmed that the herring spread into the Norwegian Sea after spawning and concentrate in the most productive areas. The locations of these depend on the hydrographical situation (the strength of the northeast-flowing warm and south-flowing cold currents). The young and adolescent herring spread through the Norwegian fjords and into the Barents Sea, and after a few years migrate westwards to mingle with the spawning stock, either for a short period in the open ocean or off the Norwegian coast. Based on scale studies, Lea (1929) concluded that the young herring must have an oceanic stage. However, he did not know about the Barents Sea distribution of the youngest age groups. Our knowledge of the whereabouts of the various age groups of the herring is more detailed today, very much due to the use of acoustic methods to survey herring in different regions. Acoustic estimates of the herring stock have been made since about 1980, first as annual surveys of the spawning grounds, but later also in the overwintering areas in Vestfjorden, Ofoten and Tysfjorden, and in the feeding areas in the Norwegian Sea. Since 1995, the herring surveys in the Norwegian Sea, where abundance estimates of herring are made, have been international and coordinated through ICES. Acoustic surveys have been the main method used for abundance estimation and stock mapping during the final decades of the 20th century.

Spawning grounds
There has been a long-term change in the importance of the different spawning sites throughout the 20th century. Some of these changes may be related to the collapse of the stock during the late 1960s and 70s, others to the influence of

environmental factors. In the first half of the last century, there were two core areas for spawning: one off Karmøy on the southwest coast of Norway, and the other off Møre on the northwest coast. Between these core areas spawning was sparse, but further north along the coast there were also important sites for spawning.

In the 1950s, when the stock decreased in abundance the importance of the southern spawning sites decreased, and spawning shifted northwards. The grounds further north near Møre gradually became more important, and by 1960 spawning on the grounds south of Bergen was negligible. Through the period 1960–1988 there were no herring spawning on the southern grounds off Karmøy (Dragesund 1970; Røttingen 1989). However, since 1989, a small component of the stock has resumed spawning on the southern spawning grounds (Johannessen *et al.* 1995).

Shifts in spawning sites and times of spawning have been observed earlier for this stock (Runnstrøm 1941) and this has been related to the size of the spawning stock and the great fluctuations in abundance (Devold 1963). The mechanisms that govern the choice of spawning sites are not yet known. Slotte and Fiksen (2000) put forward a theory that the choice of spawning site is related to the condition of the individual herring, and that the choice of spawning site is a trade-off between reaching the optimum spawning site and the use of energy to reach that site. If too much energy is used up, resorption of eggs may occur, which will not be beneficial for the spawning and survival of larvae. The distance from the over-wintering area to the spawning grounds may therefore play an important role.

Environmental causes of the choice of spawning sites are not well known. However, Runnstrøm (1941) pointed at the importance of hydrographical conditions on the spawning grounds, while Sætre *et al.* (2001) discussed the quality of spawning sites of Norwegian spring-spawning herring and came to the conclusion that sites where the bottom topography makes the water-masses around the site form eddies or retention of the water, are the most important sites. However, the mechanism that decides between main spawning areas is still not known.

During the rebuilding period, the core areas at the Møre coast were the most important spawning grounds, but spawning was also observed further north, for example on the Sklinna coast in northern Norway. As the stock increased in abundance, the spawning area increased, and by the end of the 20th century, the stock was observed to spawn at the following grounds at the Møre coast, at the Sklinna bank area, at Halten and off Lofoten. The grounds off the Møre coast were the most important. A small part of the spawning stock spawned on the grounds outside Karmøy on the southwest coast of Norway.

Feeding area
Between 1968 and 1988, when the spawning stock was small (less than about 3 million tonnes), the herring did not leave the Norwegian coast for feeding, but remained in coastal waters. Thereafter, when the spawning stock had grown to more than about 3–4 million tonnes, they migrated into the Norwegian Sea for feeding during the summer. In a historic perspective, there have been shifts in the migration pattern through the periods when the herring have migrated into the Norwegian Sea (Jakobsson and Østvedt 1999). The general rule (1907–1965),

however, was that the stock spread in the frontal areas in the Norwegian Sea and migrated to the north coast of Iceland in late summer, where there was an extensive fishery (Fridriksson and Aasen 1950; Devold 1963).

During the period when the stock was in very low abundance, between 1970 and 1986, the herring fed in coastal waters off Møre and northern Norway during the summer, and we have no indications that the stock migrated to the Norwegian Sea for feeding.

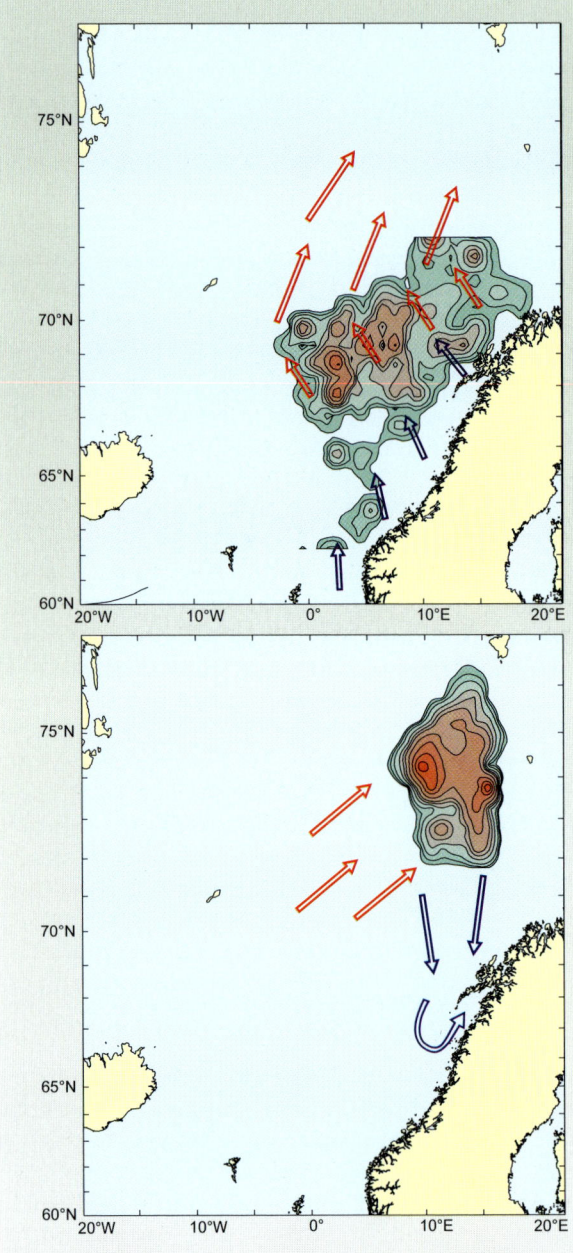

Figure 4.17
Migration pattern of the Norwegian spring-spawning herring in the early feeding season in 2000 as assessed by the ICES Planning group for pelagic surveys in the Norwegian Sea at its meeting in Tórshavn in August 2000. a) Blue arrows denote migration in April and red arrows migration in June. The distribution indicated by isolines are distributions as measured by the International survey in May. Modified after Holst et al. (2000). b) Red arrows denote migrations in July and blue arrows migrations in September. The distribution indicated by isolines are distributions as measured by the RV "G.O. Sars" in August 2000. Modified after Holst et al. (2000).

After the recovery of the stock in the early 1990s, the herring resumed the migrations into the Norwegian Sea, but has not migrated as far west as Iceland. Since 1995, annual international surveys in the Norwegian Sea in May have mapped the distribution and estimated the abundance of the herring in this area. Figure 4.17 shows the distribution of the herring in the early feeding season in 2000.

We may ask why, in the feeding season, the adult herring did not migrate to Icelandic waters in the 1980s and 1990s as they did prior to the collapse of the stock. One theory is that the herring avoided crossing the frontal area between the cold Arctic water-masses and the warmer Atlantic water-masses. Although there are examples of the herring crossing the frontal area, it seems to be a general rule that most of the herring prefer the warmer side of the front and do not cross to the colder side (Jakobsson and Østvedt 1999). During the last decades of the 20th century, the front was found further to the east than previously (Blindheim 2004). Environmental changes may therefore explain why the herring did not migrate to Icelandic waters after the recovery of the stock.

Over-wintering areas

During the recovery period through the 1970s, the spawning stock stayed close to the Norwegian coast and over-wintered in certain fjords, such as Vinjefjorden in Møre and in Gratangen in northern Norway. This distribution pattern continued throughout 1975–1986. After the recovery of the stock in the late 1980s, when the 1983 year-class recruited, the spawning stock over-wintered in a few specific fjords in northern Norway; Vestfjorden, Ofotfjorden and Tysfjorden. After feeding in the Norwegian Sea, the stock returned, during the past decade, to the Norwegian coast in September and entered these fjords somewhat later. Figure 4.18 shows echo-registrations of herring during day as recorded by the RV "Michael Sars" on 17 January 1992. In early January, the herring left the fjords and started the migration southwards along the Norwegian coast and towards the spawning grounds.

There was a sharp change in the over-wintering area after the collapse of the stock. In the period 1950–1967, the stock wintered every year in a rather small area some 200 nm east of Iceland. Then, when the stock was very low, the over-wintering area shifted to the fjords along the west coast of Norway. In the late 1980s, the fish continued to wintering in the fjords even after the stock had migrated into the Norwegian Sea to feed during the summer. The recovered stock did not spend the winter months in the same fjords as the remainder of the "old" stock had done. This stock wintered in a few specific fjords on the Møre coast and in the north of Norway.

Assessment of the stock in the period 1970–2000

In this period, two main methods were central for the monitoring of the development of the Norwegian spring-spawning herring stock; acoustic abundance estimation and tagging experiments.

Acoustic abundance estimation

In the early 1980s, acoustic abundance estimation and mapping of the distribution of herring became the most important methods for monitoring the development of the herring stock. Through surveys on the spawning grounds during the winter

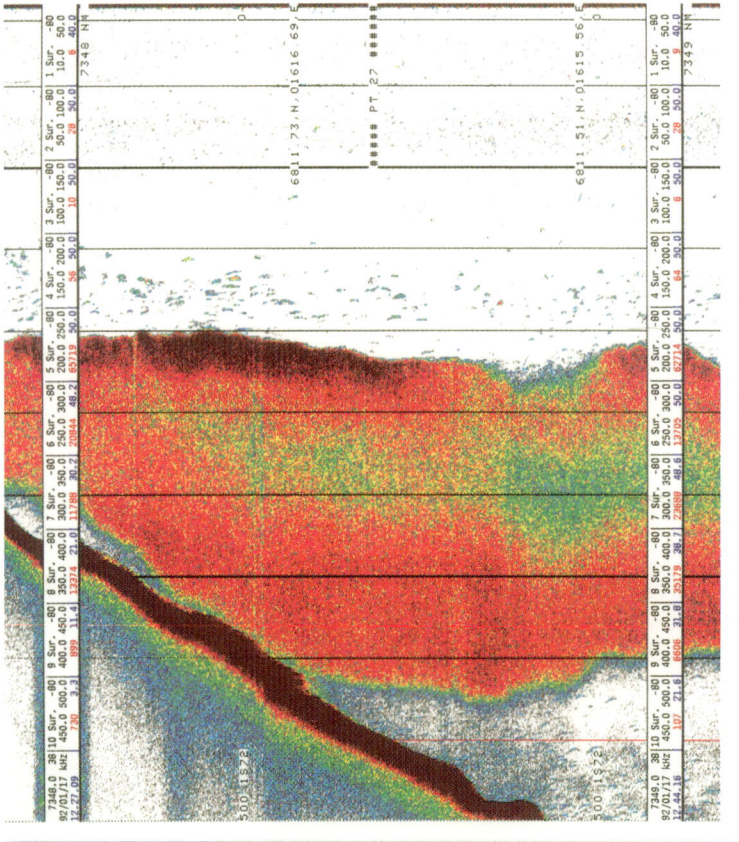

Figure 4.18
Echo-registrations of herring during daytime as recorded by the RV "Michael Sars", 17 January 1992.

and spring, the development and geographical distribution of the spawning stock were followed closely through the years of recovery. In the late autumn, acoustic surveys were carried out in fjords and coastal waters in order to assess the level of the coastal component of recruitment to the stock. When the rich 1983 year-class recruited to the spawning stock in 1988, it overwintered in a few fjords in northern Norway, Ofotfjorden and Tysfjorden. From then on, an acoustic abundance index of the spawning stock has been obtained when the stock has been located in these fjords in December and January. It is assumed that throughout the 1990s, most of the spawning stock was to be found in this specific area, and a number of *in situ* acoustic experiments were carried out to improve the acoustic abundance estimation method.

After a further increase in biomass in the mid-1990s, the spawning stock started to migrate into the Norwegian Sea for feeding migration after spawning. From 1995 on, an international coordinated acoustic survey was performed in the Norwegian Sea in order to estimate spawning stock biomass and map its geographical distribution in May–June (Figure 4.17). In some years a survey was also carried out by Norway in August for further studies of the geographical distribution. The series of abundance estimates from the coordinated acoustic survey in early summer has given the most consistent results for assessment purpose.

Tagging experiments

From 1975 on, the state of the adult stock of Norwegian spring-spawning herring has been assessed by tuning the Virtual Population Analyses (VPA) against stock estimates obtained from tagging. Both the tagging project and the model used to process the data have been described in working documents and reports of the Atlanto-Scandian Herring and Capelin Working Group (ICES 1984, 1986).

The tagging project using internal steel tags was initiated in 1975, since when herring have been tagged and released annually at various sites on the west coast in April–May. Before 1982, the herring were caught by purse seine, towed to the shore and kept in keep nets before tagging. Since 1982 they have been brailed from the seine to tanks on board the seiner and tagged and released from the tanks. The tagged herring are released in batches and under various conditions, and mortality due to tagging is expected to be variable.

The tagged herring are recovered by screening herring catches using a specially designed internal tag detector. The efficiency of the detector is tested by mixing tagged herring with the catch before screening.

The recoveries used for stock assessment are all obtained from winter catches of spawners and pre-spawners. The commercial herring winter fishery was prohibited until 1984, but experimental fishing for tag recovery was allowed during the winter between 1977 and 1983. The commercial winter fishery was opened in 1984, and in subsequent years both commercial and experimental catches have been screened for tags. The herring stock has developed in two separate units, and the data on releases and corresponding recoveries are separately processed by the two stock components. Tagging is carried out during the feeding season, and in areas where shoals from both components are found. It has therefore been difficult to allocate the releases to components when released. The tagged herring are released in batches of 2 000–10 000 individuals, and the allocation of the batches to components is made on the basis of the recoveries, i.e., the position and the age-composition of the catches from which the bulk of the recoveries are retained. The boundary between the spawning grounds of the two components runs at about 63°N.

Analytical assessments

In 1978, the first cohort analysis (VPA) of Norwegian spring-spawning herring was carried out (Dragesund and Ulltang 1978), covering the period of 1950–1971. This was a turning point for assessments of Atlanto-Scandian herring. From then on, analytical analyses have been performed annually by the assessment working group dealing with Norwegian spring-spawning herring, and it has become routine to carry out VPAs (ICES 1983; ICES 2000). In the ICES symposium 1978: The Assessment and Management of Pelagic Fish Stocks, Dragesund, Hamre and Ulltang (1980) made a comprehensive contribution to the biology and population dynamics of Norwegian spring-spawning herring. Their paper offers a description of the changes and the causes underlying the distribution and abundance of the stock from 1950 to 1978. The paper includes simulations of possible developments of the stock on the basis of various conservation strategies through the 1960s, when the stock collapsed. One such strategy simulated the likely effects of closing the small-herring fishery (0- and 1-year-olds) from 1960 onwards. Fishing mortality for older age groups was set equal to the VPA values. According to this simulation, the spawning stock biomass would not collapse. It would reach a minimum level of about 1.6 million tonnes in 1970, compared to the estimate of 60 000 tonnes produced by the VPA at the time. The simulation also indicated that the total yield of adult herring through the 1960s would have been substantially higher if the exploitation of 0- and 1-year-old fish had been prohibited.

An estimate of the stock trajectory of Norwegian spring-spawning herring throughout the 20th century, and its relation to environmental data, was made by Toresen and Østvedt (2000), who analysed stock size fluctuations for the period 1907–1998. In general, the analytical assessments produced by the VPAs and their later derivatives reveal inconsistencies in the data employed (for example catch data), which have made scientists more aware of their quality and variability. The use of analytical assessments has helped the scientists working on fish population dynamics to understand how the structural features in a stock vary and interact.

Since the beginning of the 1970s, the Norwegian spring-spawning herring has annually been assessed by the Atlanto-Scandian herring assessment working group in ICES. The analytical assessments found that the spawning stock slowly recovered to a level of about 600 000 tonnes in 1983. This is a very low level for this stock, and far below levels at which it is expected to produce rich year-classes. The time series show that when this stock has a biomass below 2.5 million tonnes, it has a very low chance of producing rich year-classes. Therefore, it was a surprise when the stock produced a very rich year-class in 1983. There must have been outstandingly good environmental conditions for the survival of fish larvae that year, because the Northeast Atlantic cod stock also produced a very rich year-class. As 0-group, the herring spread in the Barents Sea, and the year-class was recorded as being the most abundant year-class in the history of that survey (starting in 1965). The year-class recruited to the spawning stock in 1987–1989, and raised the spawning stock to a level of about 4 million tonnes in 1989.

In 1991 and 1992 recruitment was successful, and by the end of the 1990s, the stock had increased to about 10 million tonnes when these year-classes fully recruited to the spawning stock (ICES 2003).

Management in the rebuilding phase

In 1977, coastal states acquired greater influence over their "own" resources. The implementation of the 200-mile exclusive economic zones (EEZs) confirmed their ownership of resources and made it much easier to implement regulatory measures. The size of the international areas containing large fishery resources decreased drastically, and the need for multilateral agreements suddenly became more important.

Norwegian spring-spawning herring were located within the Norwegian EEZ from 1970 onwards (Institute of Marine Research 1995). The stock was at low levels of abundance (ICES 1999) and was found only in inshore waters along the Norwegian coast (Dragesund *et al.* 1997). The management of the stock was therefore regarded by Norwegian authorities as a national matter. The goal of management was to rebuild the spawning stock as rapidly as possible to a level above the Minimum Biological Acceptable Level (MBAL) of 2.5 million tonnes. The fishery was therefore strictly regulated, with a minimum landing size of 25 cm (from 1977 on) and the lowest possible catch of adult herring through

1976–1980. Through 1972–1982, less than 20 000 tonnes of herring a year were landed, except in 1977. Throughout the late 1970s and early 1980s, there was a "struggle" between the scientists at the Institute of Marine Research in Bergen (as advisors) on the one hand, and the fishing industry on the other, concerning how much herring could be harvested. The fishermen found herring on the coast all year round, and availability was occasionally relatively good. However, in 1979, the Institute produced an abundance estimate of the spawning stock, based on tagging experiment data. This estimate showed that the stock was still at a very low level, and served as a basis for the continuation of the strict regulations (ICES 1979; Hamre 1990). ICES therefore continued its advice for zero catches (begun in the 1970s) until 1984. After 1980 the fishery was expanded somewhat. After intense pressure from the Norwegian fishing industry, the authorities agreed on small annual quotas, limited to a maximum yield corresponding to a fishing mortality rate of not more than 0.05.

In 1984, when the strength of the 1983-year-class was known, the purse seiners were allocated a small quota on the spawning grounds, which was a turning point. From 1987 onwards, the USSR was allocated quotas of about 10% of the total set by Norway, but allocated to the USSR in the Joint Norwegian-USSR Fisheries Commission. This was done reluctantly by the Norwegian authorities and after a demonstration fishery of some 27 000 tonnes of juvenile herring carried out by the Soviet fleet in the Barents Sea in 1986. USSR thereafter joined in the management of the stock. By the early 1990s, the spawning stock was increasing (ICES 1999) and started to migrate into the Norwegian Sea during the summer. Accordingly, the stock again became a "hot" issue in the new NEAFC (North-East Atlantic Fisheries Commission). The organisation had always discussed the herring stock at its annual meetings, but when it again appeared in international waters and more nations expressed an interest in it, NEAFC suddenly had to play a more active role.

In 1995, an international scientific working group with participation from Iceland, Russia, the Faroe Islands and Norway was established to assess the zonal attachment of Norwegian spring-spawning herring. A report from the group (Anon. 1995) was completed in 1995. On the basis of the results received from the Working Group, NEAFC agreed in 1996 to set a TAC (total allowable catch) for the stock and to allocate this TAC among the five coastal states: Iceland, Faroe Islands, Russia, Norway and the European Union. For 1997, NEAFC decided on an additional catch limit in international waters in the Norwegian Sea. The five coastal states that had made the agreement on the TAC, as well as Poland, were given the right to fish this extra quota. In 1999, the five coastal states agreed on a long-term management regime for Norwegian spring-spawning herring, implementing the precautionary principle. In 2001, this management plan was further developed, and its main elements are to 1) maintain a level of spawning stock biomass (SSB) greater than the critical level of 2.5 million tonnes (B_{lim}), 2) to restrict fishing on the basis of a TAC consistent with a fishing mortality rate of less than 0.125 for the relevant age groups, and 3) should the SSB fall below a reference point of 5 million tonnes (B_{pa}), to adapt fishing mortality rate in the light of scientific estimates of the conditions in order to ensure rapid recovery of the SSB to a level in excess of 5 million tonnes. The basis for such an adaptation should be at least a linear reduction in the fishing mortality rate from 0.125 at B_{pa} to 0.05 at B_{lim}.

The ecological role of herring in the Northeast Atlantic

The Norwegian spring-spawning herring plays a key role in ecosystems in the Northeast Atlantic Ocean. The adult stock feeds on plankton, mainly copepods, in the Norwegian Sea and transforms energy from the plankton organisms into gonads. The gonads are spent during spawning along the Norwegian coast in February/March each year. The weight of the gonads equals about 25% of the weight of an individual herring. Hence, if the spawning stock is 6 million metric tonnes, the total weight of the spent gonad products will be about 1,5 million tonnes. These products, at first in the form of fertilized eggs, and thereafter larvae and juvenile fish, enrich the ecosystems along the Norwegian coast and the Barents Sea in the form of huge quantities of food particles. Some of this biomass will end up as a new year-class of herring, but most of it will be important food for fish and other animals living along the coast of Norway. After hatching, the herring larvae drift with the Norwegian coastal current and end up in the Barents Sea as juveniles in the late summer of the year they were born. Here they spend the first three or four years of their life before they migrate westwards and into the Norwegian Sea to join the adults.

Throughout the 1980s, an ecological comprehension of the interactions between the main fish stocks in the Northeast Atlantic arose. In the Barents Sea, it was found that young herring and capelin interact. When Norwegian spring-spawning herring produce rich year-classes they occupy the Barents Sea, and they spend their first three or four years in the area. In the southern part of the Barents Sea the distribution of small herring overlaps the distribution of capelin larvae. This overlap highly affects the survival of capelin fry because the young herring feed on the larvae. When young herring are present in the sea, capelin recruitment nearly collapses. The interactions between young herring and capelin larvae are confirmed by the time series of acoustic biomass estimates of capelin in the area, which show a severe drop in the capelin stock whenever herring are also there. Field studies have shown that young herring feed on capelin larvae. Gjøsæter and Bogstad (1998) estimated the effects of the presence of herring on the stock-recruitment relationship of Barents Sea capelin and found a significantly better fit when the years when young herring were present are excluded.

When herring are present and influence the capelin stock, the whole Barents Sea ecosystem is affected because there is a rather close relationship between the abundance of capelin and the condition of Northeast Arctic cod. The condition of cod directly influences its spawning efficiency, and thus the population dynamics of the most important predator in the area (Barros *et al.* 1998). Hamre and Hatlebakk (1998) described these relationships and developed a model in which oceanographic conditions play a major role in the dynamics of the Barents Sea system. According to the model, when winter temperature is high on the Norwegian coast, there is a higher probability of production of rich year-classes of herring (Toresen and Østvedt 2000). Fluctuations in temperature occur at intervals of around 8–12 years, and the dynamics of the ocean and its living resources are therefore affected periodically. The Norwegian spring-spawning herring is therefore looked upon as a key species in the Northeast Atlantic, affecting the abundance and dynamics of several stocks and their interactions.

REFERENCES

Anon. 1995. Report of the Scientific Working Group on Zonal Attachment of Norwegian spring-spawning herring. Chaired by Ingolf Røttingen, Institute of Marine Research, Bergen, Norway. 95/624-14-431. 47 pp. Unpublished.

Aksnes, D.L., Blindheim, J. 1996. Circulation patterns in the North Atlantic and possible impact on population dynamics of Calanus finmarchicus. Ophelia, 44, 7–28.

Barros, P., Tirasin, E.M., Toresen, R. 1998. Relevance of cod (Gadus morhua L.) predation for inter-cohort variability in mortality of juvenile Norwegian spring-spawning herring (Clupea harengus L.). ICES Journal of Marine Science, 55: 454–466.

Beverton, R.J.H., Hylen, A., Østvedt, O.J., Alvsvåg, J., Iles, T.C. 2004. Growth, maturation and longevity of maturation cohorts of Norwegian spring-spawning herring. ICES Journal of Marine Science, 61: 165–175.

Blindheim, J. 2004. Oceanography and climate. In: Skjoldal, H.R. (Ed.): The Norwegian Sea Ecosystem, Tapir Academic Press, Trondheim, 2004. ISBN; 82-519-1841-3.

Boeck, A. 1871. Om silden og sildefiskerierne navnlig om det norske Vaarsildfiske. I [eneste] Indberetning til den kgl. Norske Regjerings Departement for det Indre om foretagne praktisk-videnskabelige Undersøkelser. Trykt efter Foranstaltning af det kgl. Departement for det Indre, Christiania.

Broch, H. 1906. Foreløpige meddelelser om sildeundersøgelserne, ÅFN 1905, Bergen, s. 442–451.

Cushing, D.H., Burd, A.C. 1957. On the herring of Southern North Sea. Fisheries Investigations, London, 20(11): 1–31.

Dahl, K. 1907. The scales of the herring as a means of determining age, growth and migration. Report on Norwegian Fishery and Marine Investigations, 2(6). 36 pp.

Devold, F. 1952. A contribution to the Study of the Migrations of the Atlanto-Scandian Herring. Rapports et procès-Verbaux des Réunions de Conseil Permanent International pour L'Exploration de la Mer, 131 (13): 103–107.

Devold, F. 1953. Tokter med "G.O. Sars" i Norskehavet vinteren 1952/53. Fiskeridirektoratets småskrifter. 1953: 6, 19 pp.

Devold, F. 1963. The life history of the Atlanto-Scandian herring. Rapports et procès-Verbaux des Réunions de Conseil Permanent International pour L'Exploration de la Mer. 154 (4): 98–108.

Dragesund, O. 1970a. Factors influencing year-class strength of Norwegian spring-spawning herring (Clupea harengus Linné). Fiskeridirektoratets Skrifter, Serie Havundersøkelser 15: 381–450.

Dragesund, O. 1970b. Distribution, abundance and mortality of young and adolescent Norwegian spring-spawning herring (Clupea harengus Linné) in relation to subsequent year-class strength. Fiskeridirektoratets Skrifter, Serie Havundersøkelser, 15: 451–556.

Dragesund, O., Johannessen, A., Ulltang, Ø. 1997. Variation in migration and abundance of Norwegian spring-spawning herring (Clupea harengus L.). Sarsia, 82: 97–105.

Dragesund, O., Ulltang, Ø. 1978. Stock size fluctuations and rate of exploitation of the Norwegian spring-spawning herring, 1950–1974. Fiskeridirektoratets Skrifter, Serie Havundersøkelser, 16: 315–337.

Dragesund, O., Jakobsson, J. 1963. Stock strength and rates of mortality of the Norwegian spring-spawners as indicated by tagging experiments in Icelandic waters. Rapports et Procès-Verbaux des Réunions du Conseil International pour l'Exploration de la Mer, 154: 83–90.

Dragesund, O., Hamre, J., Ulltang, Ø. 1980. Biology and population dynamics of the Norwegian spring-spawning herring. Rapports et Procès-Verbaux des Réunions de Conseil Permanent International pour L'Exploration de la Mer, 177: 43–71.

Einarsson, H. 1952. On parallelism in the year-class strength of seasonal races of Icelandic herring and its significance. Rapports et Procès-Verbaux des Réunions de Conseil Permanent International pour L'Exploration de la Mer, 131: 63–70.

Fossum, P., Høines, Å., Røttingen, I., Slotte, A., Torstensen, E., Sund, E. 1997. Norsk vårgytende sild i området sør for 62°N. Utredning for Fiskeridirektoratet som grunnlag for regulering av NVG-sild sør for 62°N. 33 s.

Fridriksson, A. 1944. "Nordurlands-Síldin" (The herring of the North Coast of Iceland). Atvinnudeild Háskólans, Rit Fiskideildar 1944, Nr. 1.

Fridriksson, Á., Aasen, O. 1950. The Norwegian-Icelandic herring tagging experiments. Report No. 1. Fiskeridirektoratets Skrifter, Serie Havundersøkelser. 9(11): 1–43.

Gjøsæter, H., Bogstad, B. 1998. Effects of the presence of herring (Clupea harengus) on the stock-recruitment relationship of Barents Sea capelin (*Mallotus villosus*). Fisheries Research 38: 57–71.

Hamre, J. 1989. Life History and Exploitation of the Norwegian spring-spawning herring. Proceedings of the fourth Soviet-Norwegian Symposium, (ed. Terje Monstad). Bergen, 12–16 June 1989.

Hamre, J. 1990. Life history and exploitation of the Norwegian spring-spawning herring. Proceedings of the fourth Soviet-Norwegian Symposium. Ed. T. Monstad. Biology and Fisheries of the Norwegian spring-spawning herring and blue whiting in the northeast Atlantic. Bergen, 12–16 June 1989. p 5–41.

Hamre, J., Hatlebakk, E. 1998. System Model (Systmod) for the Norwegian Sea and the Barents Sea. In: Models for multispecies management (eds T. Rødseth). Physica-Verlag, Heidelberg, pp. 94–115.

Hjort, J. 1914. Fluctuations in the great fisheries of the Northern Europe viewed in the light of biological research. Rapports et Procès-Verbaux des Réunions de Conseil Permanent International pour L'Exploration de la Mer. 20: 1–228.

Hjort, J., Lea, E. 1911. Some results of the International Herring-Investigations 1907–11. Publications de Circumstance No. 61.

Hoffbauer, C. 1898. Die Altersbestimmung des Karpfens an seiner Schuppe, Allgemeine Fishereizeitung, 19. Germany.

Holst, J.C. 1996. Long term trends in growth and recruitment pattern of the Norwegian spring-spawning herring (*Clupea harengus* Linnaeus 1758). Dr. Scient thesis. Department of Fisheries and Marine Biology, University of Bergen, Norway 1996.

Holst, J.C., Røttingen, J., Melle, W. 2004. The Herring. In: Skjoldal, H.R. (ed.): The Norwegian Sea Ecosystem. Tapir Academic Press. Trondheim, 2004.

Høglund, H. 1959. Hans Høglund tror ej på Finn Devolds teorier. Svenska västkustfiskaren. Gøteborg. 110.

ICES 1930. Fluctuations in the abundance of the various year classes of food fishes. Rapports et Procès-Verbaux des Réunions de Conseil Permanent International pour L'Exploration de la Mer. 65: 188 pp.

ICES 1984. Report of the Atlanto-Scandian Herring and Capelin Working Group. Copenhagen, ICES Headquarters 25–28 October 1985. ICES CM 1983/Assess:4.

ICES 1986. Report of the Atlanto-Scandian Herring and Capelin Working Group. Copenhagen, ICES Headquarters, 29 October–1 November 1985. ICES CM 1986/Assess: 7.

ICES 1963. Report of the Atlanto-Scandian Herring Working Group, Bergen, 22–26 April 1963. ICES CM 1963/Herring Committee No 70. 18 pp.

ICES 1964. Report of the Atlanto-Scandian Herring Working Group. ICES CM 1964/ Herring Committee, 8. 22 pp.

ICES 1965. Report of the Third Meeting of the Atlanto-Scandian Herring Working Group. ICES CM 1965, Herring Committee No 19. 39 pp.

ICES 1966. Report of the Liaison Committee of ICES to the North-East Atlantic Fisheries Commission, 1965. ICES Cooperative Research Report, Series B, 187 pp.

ICES 1979. Report of the Working Group on Atlanto-Scandian Herring. Copenhagen, 21– 23 May 1979. ICES CM 1979/H:8, 44 pp.

ICES 1999. Report of the Northern Pelagic and Blue Whiting Fisheries Working Group. ICES Headquarters, Copenhagen, Denmark, 27 April–5 May 1999. ICES CM 1999/ACFM:18, 238 pp.

ICES 2000. Report of the Northern Pelagic and Blue Whiting Fisheries Working Group. ICES Headquarters 26 April–4 May 2000. ICES CM 2000/ACFM:16. 302 pp.

ICES 2001. Report of the Northern Pelagic and Blue Whiting Fisheries Working Group. Reykjavik, Iceland 18 April–27 April 2001. ICES CM 2001/ACFM:17. 243 pp.

ICES 2003. Report of the Northern Pelagic and Blue Whiting Fisheries Working Group. ICES Headquarters 26 April–4 May 2003. ICES CM 2003/ACFM:16. 302 pp.

ICES 2004. Report of the Northern Pelagic and Blue Whiting Fisheries Working Group. ICES Headquarters 27 April–4 May 2004. ICES CM 2004/ACFM:24. 305 pp.

Jakobsson, J., Østvedt, O.J. 1999. A review of joint investigations on the distribution of herring in the Norwegian and Iceland Seas, 1950–1970. Rit Fiskideilar, 16: 209–238.

Jakobsson, J.,1980. Exploitation of the Icelandic spring- and summer-spawning herring in relation to fisheries management, 1947–1977. Rapports et Procès–Verbaux des Réunions du Conseil International pour l'Exploration de la Mer, 177:23–42

Jacobsson, J., Jónsson, E., Gudmundsdottir, A. 1996. The North Icelandic Herring Fishery and the Atlanto-Scandian Herring 1939–1969. ICES CM 1996/H:30.

Johansen, A.C. 1919. On the large Spring-spawning herring in the

North-west European waters. Medd. fra Meddelelser fra Komité for Havundersøgelser, Serie Fiskeri, Bind V, No 1, 1920.

Johansen, A.C. 1926. Investigations on Icelandic herrings in 1924 and 1925. Rapports et Procès-Verbaux des Réunions de Conseil Permanent International pour L'Exploration de la Mer. 39: 114–138.

Johannesen A., Slotte A., Bergstad O.A., Dragesund O., Røttingen I. 1995a. Reappearance of Norwegian spring-spawning herring (Clupea harengus L.) at the spawning grounds off south-western Norway. In: Skjoldal H.R., Hopkins C., Erikstad K.E., Leinaas H.P., editors Ecology of fjords and coastal waters. Amsterdam: Elsevier. p 347–363.

Johannesen A.G., Blom G., Folkvord A., Svendsen H. 1995b. The effect of local wind on the distribution of Norwegian spring-spawning herring (Clupea harengus L.) larvae. In: Skjoldal H.R., Hopkins C., Erikstad K.E., Leinaas H.P., editors. Ecology of fjords and coastal waters. Amsterdam: Elsevier. p 365–384.

Lea, E. 1924. Angaaende spørsmaalet om fredning av de yngste sild (About the question of protection of the youngest herring). Aarsberetning vedkommende Norges Fiskerier for 1924, (Annual report on the Norwegian Fisheries for 1924) No 1, 1924, s 409–426. (In Norwegian).

Lea, E. 1929a. The herring scale as a certificate of origin. Rapports et Procès-Verbaux des Réunions de Conseil Permanent International pour L'Exploration de la Mer. 54: 21–34.

Lea, E. 1929b. The Oceanic Stage in the Life History of the Norwegian Herring. Journal du Conseil., Vol. IV, No 1.

Lea, E. 1930. Mortality in the tribe of Norwegian herring. Rapports et Procès-Verbaux des Réunions de Conseil International pour l'Exploration de la Mer, 65: 100–117.

Liamin, K.A. 1959. Investigations into the life cycle of summer-spawning herring of Iceland. Special Scientific Reports U.S. Fish and Wildlife Service, 327 (1959): 166–202.

Marty, J.J., Fedorov, S.S. 1963. Features of the population dynamics of sea herring as seen from the Atlanto-Scandian stock. Rapports et Procès-Verbaux des Réunions de Conseil International pour l'Exploration de la Mer, 154: 91–98.

Nakken, O. 1998. Past, present and future exploitation and management of marine resources in the Barents Sea and adjacent areas. Fisheries Research 37 (1998): 23–35.

Runnstrøm, S. 1936. A study on the life history and migrations of the Norwegian spring-herring based on the analysis of the winter rings and summer zones of the scale. Fiskeridirektoratets Skrifter, Serie Havundersøkelser. (Report on Norwegian Fishery and Marine Investigations) Vol. V, No. 2: 5–102.

Runnstrøm, S. 1941. Racial analysis of the herring in Norwegian waters. Fiskeridirektoratets Skrifter, Serie Havundersøkelser. 6 (7): 5–10.

Røttingen, I. 1989. Reappearance of Norwegian spring-spawning herring on the spawning grounds south of 60°N. ICES CM 1989/H: 22:8.

Sars, G.O. 1879. Indberetninger til departementet for det indre om de av ham i årene 1864-78 anstillede undersøkelser angaaende saltvannsfiskeriene. Berg og Ellefsens Bogtrykkeri, Christiania: 221 pp. (In Norwegian). Christiania.

Schwach, V. 2001. Havet, fisken og vitenskapen. Fra fiskeriundersøkelser til havforskningsinstitutt 1860–2000. (The oceans, the fish and the science. From fishery investigations to an institute of marine research 1860–2000). The history of the Institute of Marine Research in Bergen, Norway. ISBN 82-7461-051-2. [In Norwegian]. 405 pp.

Seliverstova, E.I. 1970. Comparative Characteristics of the Atlanto-Scandian Herring of the 1950 and 1959 Year-Classes (Ratio of types of Growth: a Rate of Sexual Maturity). ICES CM 1970/H:21. [Mimeo.]

Slotte, A., Fiksen, Ø. 2000. State-dependent spawning migration in the Norwegian spring-spawning herring (Clupea harengus L.). Journal of Fish Biology 56: 138–162.

Smitt, F.A. 1895. Skandinaviens Fiskar, Stockholm, Sweden.

Sundby, S. 1994. The influence of bio-physical processes on fish recruitment in an Arctic-boreal ecosystem. Dr.Philos thesis, University in Bergen, Norway. 190 pp.

Sætersdal, G. 1980. A review of past management of some pelagic stocks and its effectiveness. Rapports et Procès-Verbaux des Réunions de Conseil Permanent International pour L'Exploration de la Mer 177: 505–512.

Sætre, R., Toresen, R., Søiland H., Fossum, P. 2001. The Norwegian spring-spawning herring – spawning, larval drift and larval retention. Sarsia 87: 167–178.

Toresen, R. 1990. Long-term changes in growth of Norwegian spring-spawning herring. Journal du Conseil International pour l'Exploration de la Mer, 47: 48–56.

Toresen, R., Østvedt, O.J. 2000. Variation in abundance of Norwegian spring-spawning herring (*Clupea harengus*, Clupeidae) throughout the 20th century and the influence of climatic fluctuations. Fish and Fisheries, 2000, 1, 231–256.

Toresen, R., Østvedt, O.J. 2002. Stock structure of Norwegian spring-spawning herring: historical background and recent apprehension. ICES Marine Science Symposia, 215: 532–542.

Vollan, O. 1971. Sildefisket Gjennom Tusen År. (The herring fishery through thousand years). Norsk Kulturarv (Norwegian cultural heritage). Det Norske Samlaget, Oslo. 139 pp. (In Norwegian).

Østvedt, O.J. 1963. Catch, effort and composition of the Norwegian winter herring fishery. Rapports et Procès-Verbaux des Réunions de Conseil Permanent International pour L'Exploration de la Mer 154: 109–117.

Østvedt, O.J. 1958. Some considerations concerning the homogeneity of the Atlanto-Scandian herring. Rapports et Procès-Verbaux des Réunions de Conseil Permanent International pour L'Exploration de la Mer 143: 53–57.

Østvedt, O.J. 1964. Growth and maturation of the Norwegian herring. ICES CM 1964 Herring Committee No 141. 10 pp. [Mimeo.].

CHAPTER 5

Northeast Arctic cod: fisheries, life history, stock fluctuations and management

Arvid Hylen, Odd Nakken and Kjell Nedreaas

5.1 Introduction

The stock of Northeast Arctic cod is potentially the largest cod stock in the world. Its nursery and feeding areas are in the Barents Sea and Svalbard waters (Figure 5.1). Mature fish undertake spawning migrations southward along the coast of Norway where important fisheries for spawning cod have taken place for centuries in February–March; what are known as the "skrei" fisheries. The main spawning area is Lofoten. Immature fish make seasonal east–west and north–south migrations in the Barents Sea that are related to the cooling and warming of the water masses and the distribution of prey. At the age of 3–5 years (40–50 cm in length), when capelin become a major prey item, the immature cod follow the spawning migrations of capelin to the coasts of Finnmark and Murmansk where another traditional cod fishery takes place, the spring-cod or capelin-cod fishery, in April–May. Until about 1920, the bulk of the annual yields came from these two Norwegian coastal fisheries. Between 1920 and 1940, international

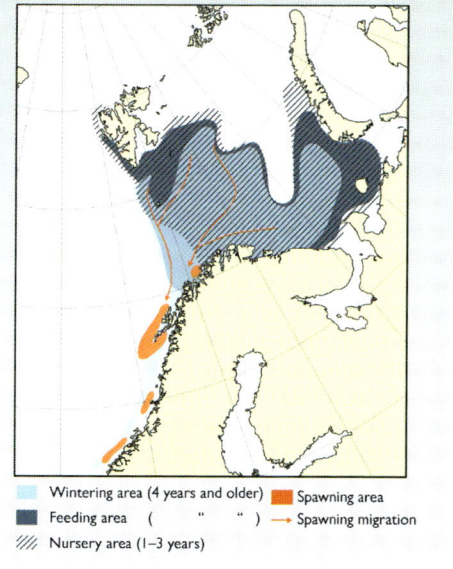

Figure 5.1
Northeast Arctic cod. Distribution and migrations (Michalsen, K. (red) 2004).

Wintering area (4 years and older)   Spawning area
Feeding area   ("   ")   → Spawning migration
Nursery area (1–3 years)

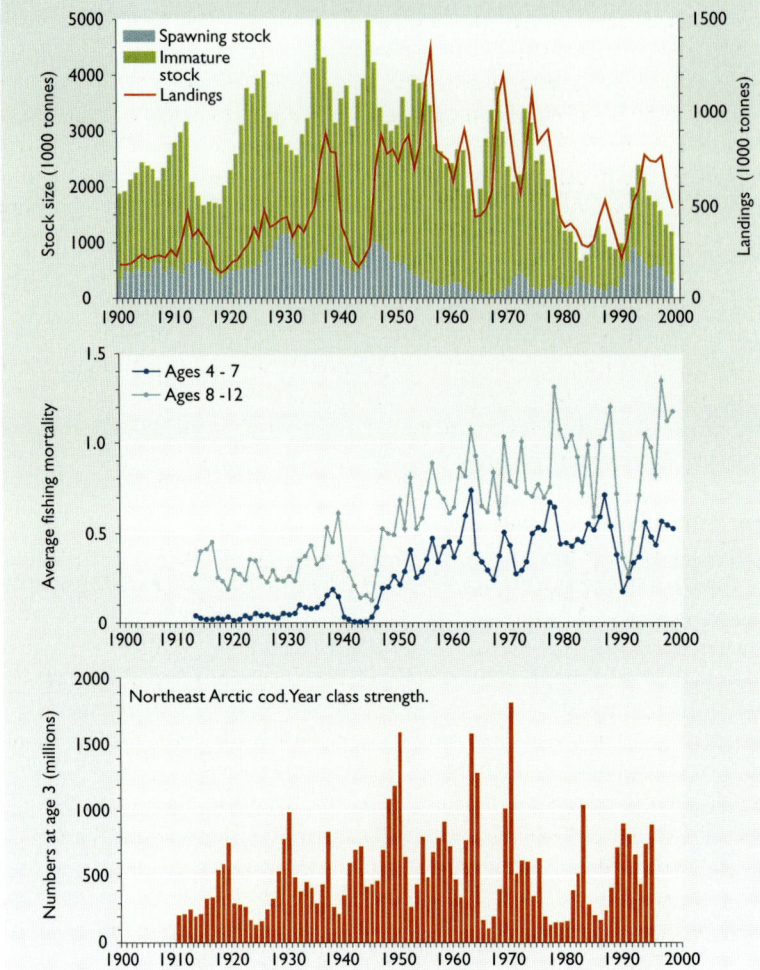

Figure 5.2
*Northeast Arctic cod.
a) Fluctuation in total stock biomass (3 years and older), spawning-stock biomass, and total catch (t).
b) Average fishing mortality (F) on ages 4–7 and 8–12.
c) Year-class abundance at age 3 (millions of fish) (Hylen 2002).*

offshore trawl fisheries developed in the Barents Sea; fisheries which, by the end of the 1930s, exceeded the total catch of cod in the traditional Norwegian fisheries. The expansion of the offshore trawl fisheries continued after a period of low fishing activity during World War II, and in the 1950s it was obvious that the large catches of young and immature fish in these fisheries were reducing yields in the skrei fisheries. Advice on fisheries management has been regularly offered by ICES since about 1960, including recommendations of a variety of measures (increased mesh size, closed areas for fishing, reduced effort, catch quotas) in order to protect small sized fish and reduce the rate of exploitation. Nevertheless, prior to the establishment of national economic zones (NEZ) in 1977, no effective management measures were ever put into operation for the fisheries on this stock.

Catches of Northeast Arctic cod have shown large temporal and geographical variations during the 20th century. Figure 5.2 shows landings, annual estimates of stock biomass and spawning stock biomass as well as fishing mortality rates and recruitment. Many authors have contributed to our present understanding of the causes of the observed fluctuations in catches, stock size and exploitation rates, and references are given in Bergstad *et al.* (1987), Nakken (1994), Sundby (2000) and Hylen (2002).

In the present chapter we summarize and evaluate the available information on trends and fluctuations in fisheries, stock size and biological characteristics for the stock throughout the 20th century. We describe the development in our understanding of the causes of these fluctuations as well as the management measures which have been advised and introduced in order to maintain the yield from the stock at a high and sustainable level.

5.2 Fisheries and yields

Detailed landing statistics for Northeast Arctic cod have been available since 1864. On the basis of information from export statistics and taxes, Øiestad (1994) presented figures for annual landings back to 1570. His results show a major increase in catches during the 19th century from a highly variable level between 5 000 and 50 000 tonnes prior to 1810 to about 200 000 tonnes at the end of the century. Øiestad also asked whether the fishery reflected the size of the stock and concluded that there was reason to believe that the main pattern of variation in catches reflected actual variations in abundance.

The development of total international landings throughout the 20th century is shown in Figure 5.3. Until the mid-1920s, the bulk of the landings was from the two traditional Norwegian near-shore fisheries; the skrei (spawning cod) fishery and the spring cod (young cod) fishery. During the 1930s, catches increased rapidly with the increasing international trawling effort in this area. Trawl catches ceased during World War II when the Barents Sea became a war zone, but increased rapidly again after the war, with total annual landings reaching a level of more than 1.3 million tonnes in 1956. Throughout the 1960s and 70s, landings fluctuated considerably due to fluctuations in the abundance of the year-classes that became available to the fishery. Annual total fishing quotas (total allowable catches, TAC) were introduced in 1975, and since 1977, Norway and Russia have shared the ownership and management responsibility for the stock (see section on management).

Landings by gear are shown in Figure 5.4. During the first two or three decades of the century, when most of the catch was from the Norwegian fisheries, nearly all the fish were caught by traditional gears; i.e. handlines, longlines and gillnets. Since the 1930s, with exception of the war years (1940–1945), 40–90 percent of the landings have been trawl catches. Trawling was not used to any

Figure 5.3
Total international landings of Northeast Arctic cod by main countries.

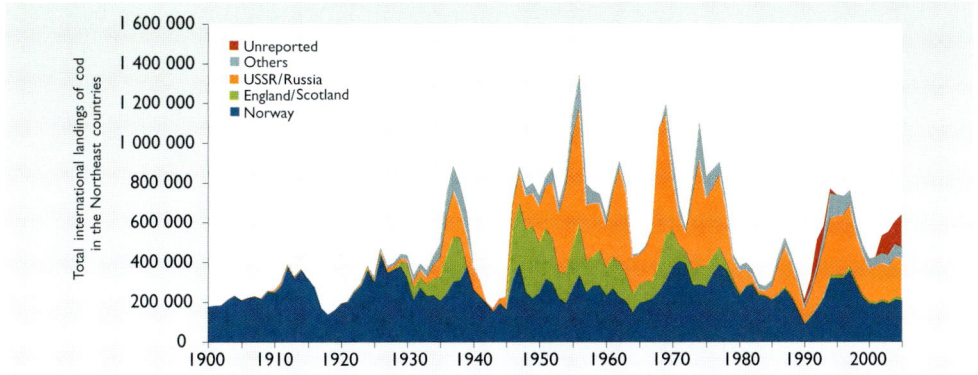

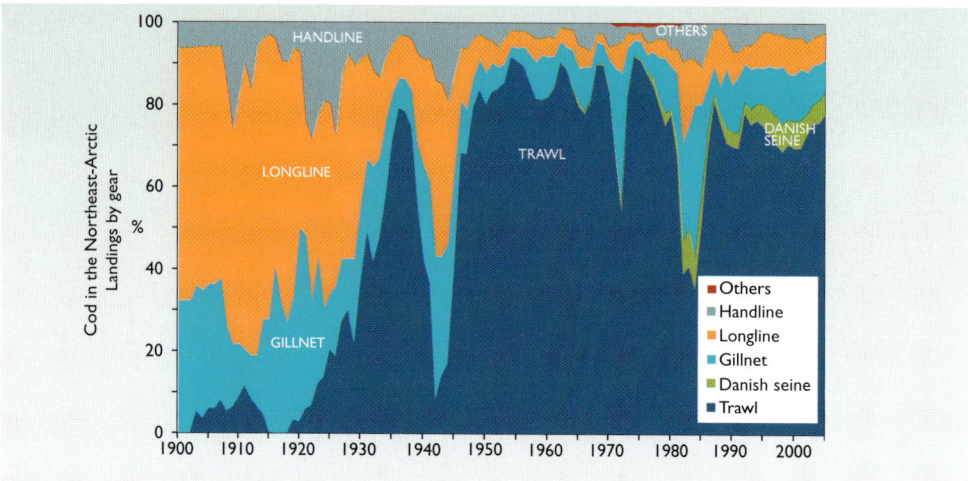

Figure 5.4
Northeast Arctic cod. Landings by gear.

extent in the Norwegian cod fisheries before the 1950s. During the pre-war decades, when other nations (Great Britain, Germany and USSR) were developing their trawl fisheries in the Barents Sea, there was a considerable resistance to trawling in Norwegian fishing communities. It was feared that offshore trawling, which captured large amounts of small and young fish, would reduce the amount of larger cod approaching the coast in winter–spring and thus lower the availability of fish to the Norwegian cod fisheries. Because of this, and in order to avoid gear conflicts between trawling and longlines/gillnets, the use of trawls in Norwegian territorial waters was forbidden in 1908, and for a long period landings of trawl-caught fish were prohibited in Norway. Before World War II, Norway had only a few (10–11) trawlers engaged in the cod fisheries. However, after the war, in the 1950s and 60s, a number of trawling licences were issued to Norwegian fishermen in order to provide the growing fillet industry with steady supplies of raw materials.

The Norwegian "skrei" fisheries
Of the different Northeast Arctic cod fisheries, the "skrei" (spawning cod) fisheries are the oldest and most important; they are as old as the history of Norway, and the development of the cod fisheries has very much been linked to the "skrei". For hundreds of years, the main fish product exported from Norway was dried cod (stockfish); i.e. cod fished during the skrei fisheries in winter–spring along the western and northern coasts, the Lofoten region being the main fishing area. Figure 5.5 shows the percentage landings by gear in the Norwegian "skrei" fisheries. Handline, longline and gillnets accounted for all catches until the late 1940s when the purse seine, trawl and Danish seine became important gears. There is a noticeable relative reduction in longline catches during the time series. In 1949–1955, the authorities permitted several hundred vessels to trial purse seine during the skrei fisheries in Lofoten, and for some years in the 1950s, the purse seine catch made up 30–40 percent of the Norwegian catch of skrei. The use of purse seining in the skrei fishery came to an end in 1959. From Figure 5.5 it appears as if no skrei were caught by trawlers in the most recent decades. This is an artefact caused by a change in the official fishery statistics in the late 1970s. Since then, only the catch of mature cod caught in the Lofoten

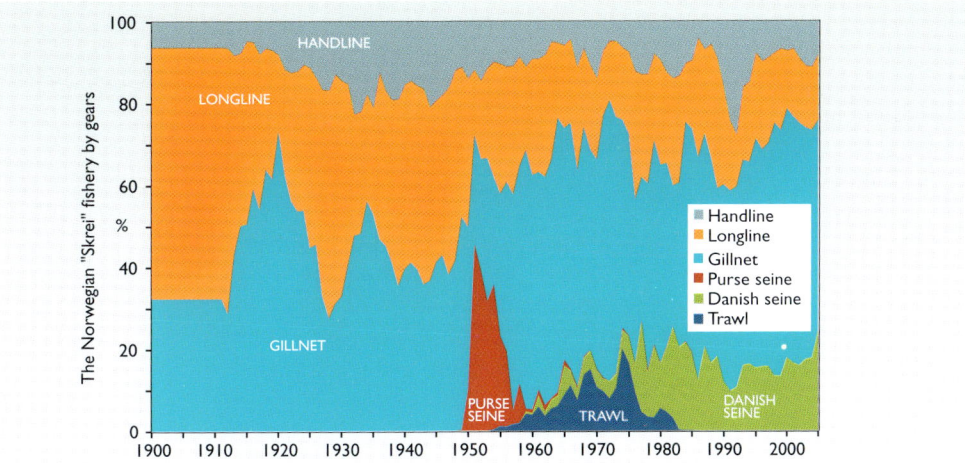

Figure 5.5
The Norwegian skrei fishery by gear. The lack of trawl catches in recent decades is due to a change in the official landing statistics.

monitoring district in the spawning season have been recorded as skrei, while in previous decades, spawning Northeast Arctic cod caught along the entire western and northern coast from Finnmark and southward were reported as skrei. The Lofoten district is in territorial waters in which trawling is not permitted. However, the southward migration pattern of the pre-spawning cod is largely outside the territorial waters in which Norwegian as well as foreign trawlers fish, and as early as the 1930s, Rollefsen (1938) thought that a substantial part of the skrei population was caught by foreign trawlers during the spawning migration. The extension of the fishery limit to 12 nautical miles in 1961 and the new management system adapted in 1977 greatly reduced foreign trawling activity along the migration route.

Figure 5.6 shows Norwegian landings of skrei in individual districts during the 20th century. Until the 1930s, substantial amounts were caught in areas south of Lofoten, and in 1912–1915, landings in the Møre–Trøndelag district exceeded those in Lofoten. South of 62°N, skrei fisheries ceased in the 1920s due to lack of fish, and in recent decades, catches in the Møre–Trøndelag district also have been very low. Along with the decline in catches in the southern districts, there has been an increase in the districts north of Lofoten. Several

Figure 5.6
The Norwegian skrei fisheries by area.

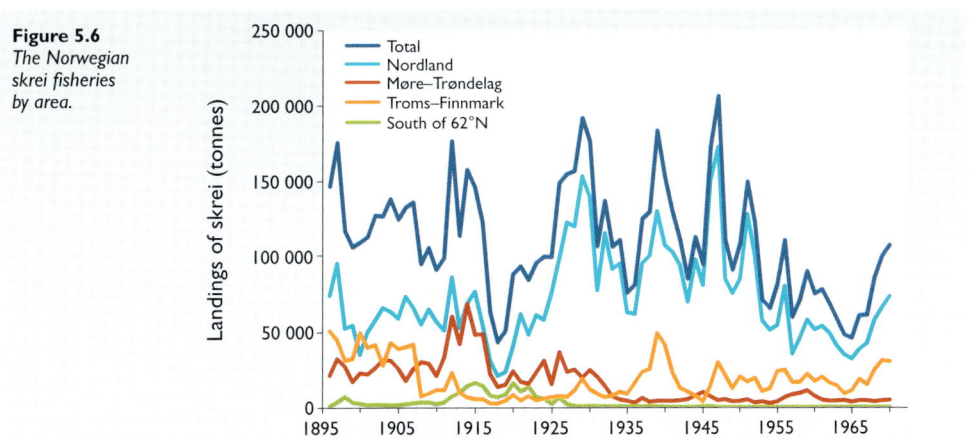

authors have interpreted this as an effect of climate change (see section 5.4). The cod fisheries attracted many fishermen, and in order to organize fishing so that gear collisions could be avoided, three regional monitoring services were established; in Lofoten and in Borgundfjord (in Møre) for the skrei fisheries, and in Troms–Finnmark during the spring cod fishery.

The Lofoten fishery
The largest and best-known of the Norwegian cod fisheries is the Lofoten skrei fishery, which has accounted for the bulk of the Norwegian catch of skrei for centuries. In the 1890s, the number of fishermen taking part in this fishery varied between 24 000 and 32 000 each year, with annual catches between 15 million and 31 million spawning cod, which gave 8 000–19 000 hectolitres of cod liver oil. Participation in the fishery as well as catches decreased in the early 20th century, but reached a new peak in the 1930s, when 22 000–32 000 fishermen and more than 10 000 vessels were present in Lofoten, catching 53 000–130 000 tonnes a year. The number of fishermen remained at about 15 000–20 000 until the mid-1950s, but then declined rapidly to about 5 000 in 1970 and to a little more than 3 000 by the end of the 20th century.

Even centuries ago, the large numbers of fishing vessels operating in a very limited area required fishing to be organized according to specific rules, and a special "Lofoten Act" was passed for this purpose. An important measure aimed at avoiding gear collisions was the division of the total fishing area into sub-areas allocated for particular types of gear, as well as fixed setting and hauling hours for gillnets and longlines so that other gears could safely operate. During the fishing season, the main services of society, (i.e. law and order, health, communications) were considerably strengthened in the villages where the fishermen lived and landed their fish. Particular houses were available for lodging; a system that was initiated by King Øistein in about 1130. Along with the modernizing of the fishing vessels and the falling participation in the fishery during the latter half of the 20th century, the need for these lodgings decreased and eventually disappeared.

The spring cod fishery
While the Lofoten fishing season was in late winter and early spring (February–March), the season for the spring cod fishery in Finnmark was somewhat later (April–June), and many cod fishermen sailed directly from Lofoten to Finnmark. At certain periods, depending on the abundance of the year-classes, catches in the Finnmark fishery exceeded those in the Lofoten fishery. A strong year-class recruits to the Finnmark fishery as 4–6-year-old fish, and thus 2–3 years before it becomes mature and appears in the Lofoten fishery. Hence a rich fishery of young cod in Finnmark was a hint that an increase in the availability of fish could be expected in Lofoten within a few years. Large temporal east–west and shallow–deep water shifts have occurred in the spring cod fishery, depending on where the capelin schools "decided" to spawn; shifts that have been linked to changes in sea temperatures (see section 5.4).

5.3 Sampling, age determinations and stock units
Sampling of individual cod from the two large seasonal fisheries for the skrei and spring cod was initiated during the first decade of the 20th century and has

been carried out annually ever since 1913. The age, size, sex and maturity stage of each fish were determined. Until 1930, age determinations were based on scale readings, a method that produced unreliable results for fish older than 7–8 years (see Hylen 2002 for a discussion of the problem). In late 1931, Rollefsen had his method ready for analysing otoliths (Rollefsen 1933, 1935). The method provided information both on the age of the fish and age at first spawning, and thus the number of times that the fish had spawned. Employing this methodology on the samples taken during the skrei fishery in 1932, Rollefsen (1933) concluded:
1. The skrei may reach maturity from their 6th to the 15th year
2. Most of them reach maturity in their 10th and 11th year
3. Skrei spawn every year after reaching maturity

Rollefsen also estimated that the number in each year-class fell by 35–40 percent for each spawning. Since 1932, age determinations have been based on otolith readings.

For centuries, fishermen have made a distinction between two main types of cod in Norwegian waters; the coastal cod that inhabit near-shore areas and fjords all the year round, and the great masses of cod entering coastal waters to spawn or feed on capelin in late winter and spring. They distinguished between the two types on the basis of the shape and colour of the fish, as did G.O. Sars when he drew them (Sars 1879). When Rollefsen studied cod otoliths (Rollefsen 1933, 1935) he found that "The otoliths of these coastal cod differ considerably in many ways from those of the skrei (Northeast Arctic cod), not only in the relative width of the zones and their structure, but even in the external form". Rollefsen also found that the two groups were characterized by differences in growth and age at first spawning; the coastal cod grows faster and reach maturation at an earlier age than Northeast Arctic cod; observations which have since been verified (Godø 1984).

Studies made by British and Russian scientists in the 1950s, revealed differences in otolith structure and zone patterns in cod caught at various locations in the Barents Sea (Trout 1955; Mankevich 1969). However, the separation into two main groups, coastal and Northeast Arctic cod, was supported by the genetic studies of Møller (1968, 1969). He suggested that the two groups were genetically distinct populations. However, Mork *et al.* (1985) found that the genetic distance between the various stocks of Atlantic cod was generally low, and they believed that this was probably a result of gene flow between stocks. More recent investigations that have compared the results of genetic studies, tagging experiments and otolith patterns, have led to the conclusion that the two groups should be regarded as separate stocks (Jakobsen 1987; Dahle 1991), and since the mid 1990s, a separate stock assessment of coastal cod has been carried out by ICES. Results from quite recent analyses support the use of otolith shape for separation of Northeast Arctic cod and Norwegian coastal cod (Stransky *et al.* 2007).

5.4 Distribution and migrations
Knowledge of the main features of the distribution and migration of the various stages and size groups of the stock was established between 1878 and 1914. Observations made during the Norwegian North-Atlantic Expedition of 1876–1878 (see Sætre 2004; Rollefsen 1966) enabled G.O. Sars to map cod

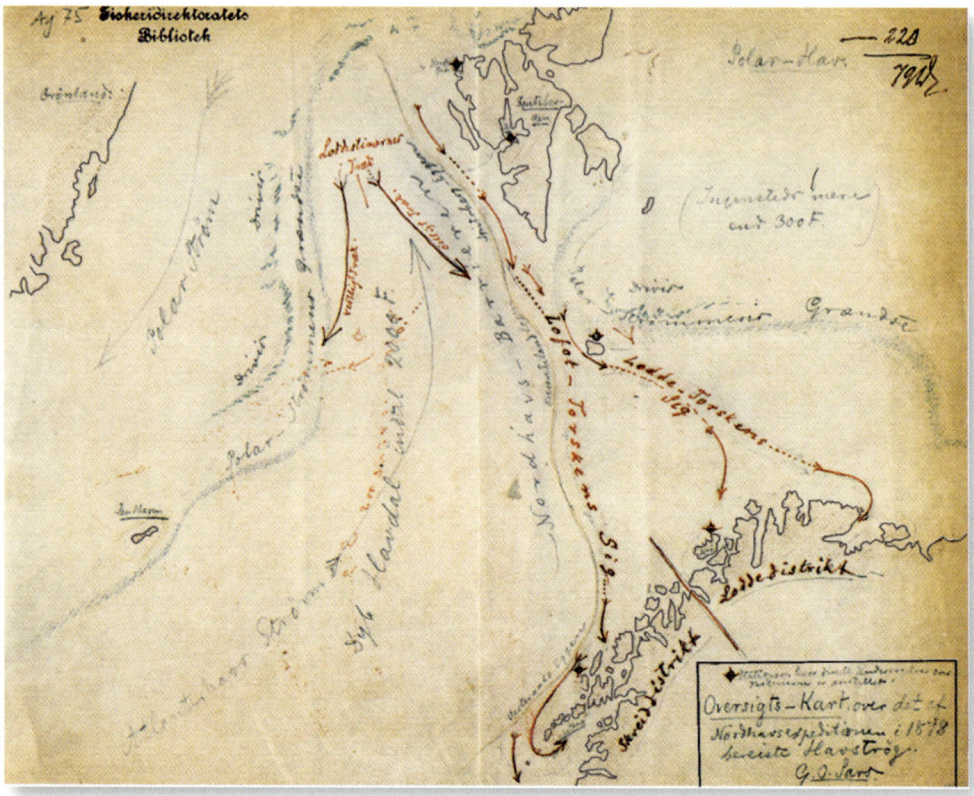

Figure 5.7
The migration of the skrei (Sars 1879).

migrations from the Svalbard banks to the coast of Norway. Extensive surveys and sampling activity by Norwegian and Russian research vessels in 1900–1903 provided information on drift patterns of eggs and larvae as well as on the seasonal changes in the spatial distribution of the young cod (Solemdal and Serebryakov 2002; Sætersdal, this volume). Large-scale tagging experiments in 1912–1913 showed that cod tagged off Finnmark made spawning migrations far south along the Norwegian coast.

Rollefsen (1966) has described G.O. Sars' discovery of the pattern of cod migration. In the mid 1860s, when Sars had found the eggs of the cod and seen its larvae hatch, he concentrated his interest on where the large quantities of cod spawning in Lofoten in winter/early spring were to be found during the rest of the year. He was able to provide an answer in the summer of 1878, when he studied samples of cod fished off Spitzbergen, near 80°N: "Sars examined the Spitzbergen cod. He found no roe or milt in it, although it was very large. He found no small fish and no cod fry along the shore. He posed the question: "Where do the Spitzbergen cod reproduce?" and writes: "Now that our last expedition has procured us a complete picture of the formation of the ocean bottom and the other physical and biological conditions in the waters we sailed, this question appears just as easy as it formerly seemed difficult. There can be no doubt that the mature individuals go south and appear along our coast as the well known Lofoten winter cod". Sars then marked the paths of migration of the Lofoten cod and the Finnmark cod on his map. While he allows the Lofoten cod to make its spawning migrations to the Norwegian coast, he holds the immature

Finnmark cod back in the Barents Sea (Figure 5.7). It was as easy as that. As a reward for his labours, Sars received a gift from his cod. Skipper Ingebrigtsen of the sloop "Amalie" of Tromsø caught a cod at Spitzbergen that had a fishhook in its jaw, with a little piece of snood still attached. The hook was sent to Sars. The snood was of a quite different type from that used at Spitzbergen, but the same type as fishermen used in Lofoten and Finnmark. The flesh of the fish had grown over the hook, which must have been there for a long time. The cod had surrendered (Rollefsen 1966).

The current map made by Mohn in 1887 based on data from the Norwegian North-Atlantic Expedition (1876–1878) (see Sætre 2004), indicated that eggs and larvae from the spawning fields at the Norwegian coast would drift northward and into the Barents Sea, and the observations made with the research vessel "Michael Sars" during 1900–1903 showed such a drift pattern; cod eggs were only observed in near-shore waters, while larvae and larger fry were found more offshore and further north (Figure 5.8). The "Michael Sars" surveys were coordinated with the work of the Russian research vessel "Andrei Pervozwanny" which started operations in the Barents Sea in 1899 (Solemdal and Serebryakov 2002). The knowledge of cod distribution gained from the extensive and coordinated survey activity in these years, in addition to the results of the tagging experiments (Figure 5.9) carried out a decade later (1912–1913), were summarized roughly as follows in the 1914 Annual Report on Norwegian Fisheries: The schools of cod which each winter/spring spawn at the Norwegian coast belong, to a large extent, to a single stock which has its feeding and nursery areas in the Barents Sea and Svalbard waters.

The main spawning fields were in the Lofoten region. However, before 1920, there were large skrei fisheries all along the western coast, and in some years the

Figure 5.8 The outer limits of the spawning products of cod. Results from cruises in 1901–1904. I Eggs, II Larvae, III Post-larvae (Hjort 1914).

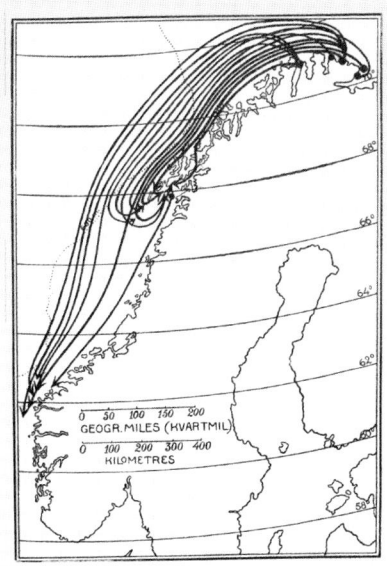

Figure 5.9 Main results of cod tagging experiments in 1912–1913 (Hjort 1914).

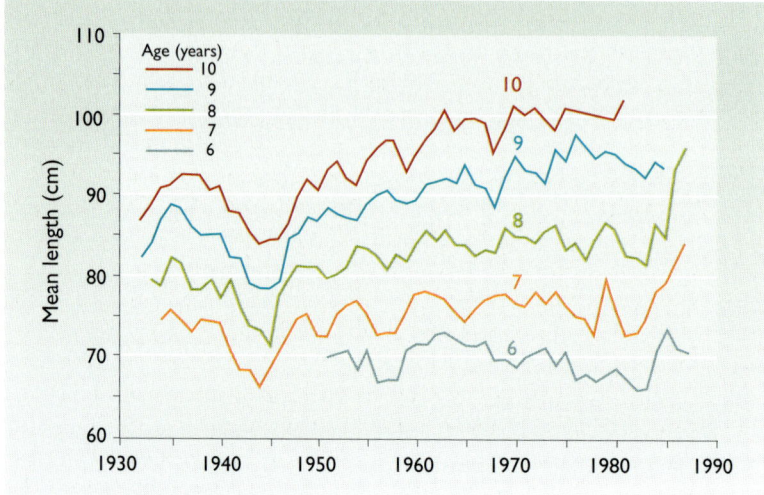

Figure 5.10
Mean length at age of first time spawners (Jørgensen 1990).

catches in the Møre region were on a level with landings in Lofoten (Figure 5.6). At the ICES annual meeting in 1926, Oscar Sund described the distribution area of the stock as follows: "As is well known, the cod population of the Norwegian coast north of Cape Lindesnes (Naze of Norway) is effectively separated from the cod populations of the other regions except that of the shallow Arctic seas north and east of Finnmark. These seas form "the grazing grounds", while the coastal waters of Norway between the northern and southern extremities of the country offer the conditions necessary to the propagation of this cod population. A statistical survey of the northern cod population, which may properly be termed the Arcto-Norwegian, should therefore include evidence both from the different parts of the coast and from the northern waters" (Sund 1927). Clearly, Sund was of the opinion that the great fisheries for pre-spawning and spawning cod which took place each year in February–March all along the coast, were mainly based on fish belonging to the stock of Northeast Arctic cod or Arcto-Norwegian cod, as it was called until the 1970s.

A number of tagging experiments (Dannevig 1951; Trout 1955; Berger 1965), increasing and more wide-ranging fishing activities and in particular the information obtained from surveys following World War II (Maslov 1956; Baranenkova 1961; Midttun and Sætersdal 1956), have verified the main picture established by Sars, Hjort and Sund and added some important details regarding the distribution of the stock. There seems to be none or little migration of immature cod between the two main nursery areas, the Bear Island–Svalbard region and the southeastern part of the Barents Sea. Immature fish feed both on the bottom and in the midwater layers and make seasonal east–west and north–south migrations. The amplitude or range of these migrations increases with age, and at the age of three to five years, when capelin become a major food item, the cod follow the spawning migrations of capelin to the coasts of northern Norway and Murmansk. Some older immature fish join the mature population on its migration towards the spawning fields further south, the so-called "dummy runs" (Trout 1957; Woodhead 1959). Tagging experiments on spawners have indicated that the majority of fish return to the same spawning field for several years in succession (Godø 1984).

In addition to seasonal displacements, temperature- and climate-related displacements have been observed on both small and large time- and space-scales. In periods of warm climate in the Barents Sea, the distribution area is extended towards the north and east, as compared to cold-climate periods, when the fish tend to concentrate in the south-western parts of the sea. In the 1870s, a rich fishery occurred during the summer for some years off western Spitsbergen. The fish disappeared from these banks in the early 1880s, probably because of an unfavourable ocean climate, and until about 1920 no fishable concentrations were found on the banks in the Bear Island–Svalbard area (Iversen 1934). During the warming of the North Atlantic and the Barents Sea in the 1920s and 30s, the area of distribution of the stock was extended substantially towards the north and east (Tåning 1949), and the banks in the Bear Island–Svalbard area, as well as those in the eastern part of the sea, near Novaya Zemlja, became important fishing grounds during summer–autumn. Simultaneously there seem to have been a change in spawning areas; catches of skrei in the areas south of Lofoten, which before 1920 had been substantial, gradually decreased during the 1920s and 30s, while rising in Finnmark. Sætersdal (1960) believed that changes in ocean climate caused displacements of the entire distribution area for the stock; in cold periods both feeding areas and spawning fields were shifted towards the south and west (i.e. towards warmer waters) compared with warm periods. His opinion was based on the shifts in the fisheries and the findings of Eggvin (1938), and is supported by more recent studies based on longer time series. These studies show that while the quantity of spawning cod fished in Lofoten always makes up a substantial portion of the total Norwegian catch of skrei, catches north and south of Lofoten have varied with sea temperature. In cold periods a larger proportion of the spawners have moved to the southern districts than in warm periods, when spawning to the north of Lofoten has increased (Godø 2003; Sundby and Nakken 2004). Large short-term shifts in the geographical distribution, lasting for a few years and closely related to changes in temperature, have also been observed (see Ottersen et al. 1998 for references). From 1977–1980, temperatures in the Barents Sea were very low, and the area of distribution of the immature cod decreased substantially (Nakken and Raknes 1987), a matter which influenced the quality of the stock assessments in those years (see page 111).

The vertical distribution of the cod also changes with temperature. The depth of the concentrations of spawning cod varies according to the depth of the transition layer between the winter-cooled coastal water and the warmer and more saline Atlantic water below. Spawning cod generally stay within this transition layer, which may rise and sink very rapidly in the course of a season, due to changes in air pressure and wind conditions. Before echo-sounders came into widespread use, fishing gave uneven results due to the vertical shifts of this layer (Sund 1933; Eggvin 1938), which led to similar shifts in fish distribution.

5.5 Growth and maturation

Both the rate of growth of Northeast Arctic cod and the age at which the fish become mature have shown large changes over time.

Long-term changes in growth rates are shown in Figure 5.10. The mean length of first-time spawners at the age of 8–10 years increased during the period after the Second World War. Short-term variations in growth have been consider-

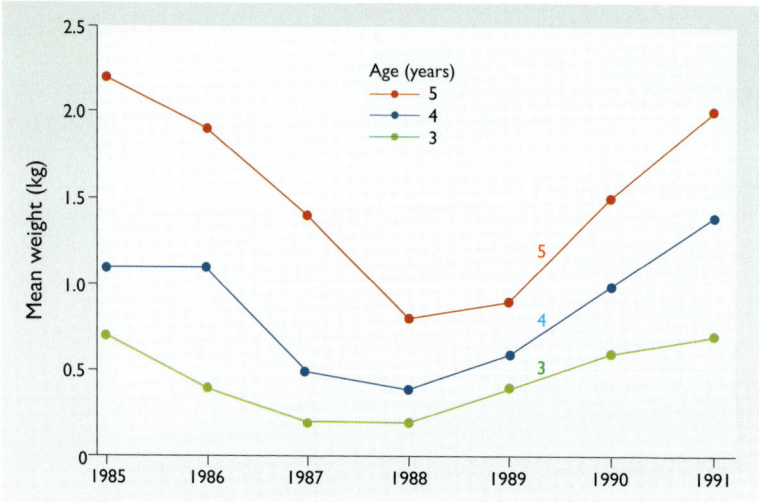

Figure 5.11
Mean weight at age during a collapse of the capelin stock (Nakken et al. 1996).

able, and particularly noticeable is the great reduction in mean length at age during the period 1938–1944, when the mean length of all age-groups from 6–10 years decreased by approximately 8–10 cm. This coincided with a complete disappearance of spawning capelin off the coast of Finnmark in 1938–1942 (Olsen 1968). Reliable data for weight at age in this period are lacking, but for 65–70 cm long fish, the length reductions are equivalent to a weight reduction of about 40 percent. A similar reduction in weight of young cod (3–5 years of age) was detected during the second half of the 1980s (Figure 5.11), when the weight of five-year-old cod fell by about 50 percent in the course of three years. Both of these periods occurred in years when the capelin stock was in a poor state. Towards the end of the 1980s, the cod ate mostly amphipods and gadoid juveniles (including cod) to compensate for the lack of capelin, and the nutrition value of such a diet is poorer than that of capelin. A large stock of small herring in the Barents Sea could probably have offset the weight reductions in cod in 1986–1988 (Hamre 1988), but the stock of small herring (the 1984 and 1985 year-classes) was rapidly eaten, and the growth and condition of cod became extremely poor. However, when the capelin stock recovered in 1989–1990, cod growth rates increased rapidly (Figure 5.11). Early in the century too there was an abrupt decline in cod growth. In his report on the cod fisheries in 1903, Johan Hjort described a situation similar to that observed in 1987; cod fisheries were extremely poor because hordes of harp seals invaded the fishing grounds and scared the fish into deep waters. The cod were so lean that their livers sank in seawater and thousands of dead seabirds were floating off the Finnmark and Murman coasts; probably starved due to lack of capelin (Hjort 1903).

Male cod reach maturity at a younger age and smaller size than females. For Northeast Arctic cod this difference is quite significant. Rollefsen (1937) found that nearly all six- and seven-year-old spawners were males, while females were dominant in the spawning stock at ages of 11 years and above. During the 1990s, the difference in age at 50 percent maturity between females and males was about one year (Ajiad et al. 1999) and the data indicated that in many years, the ratio of males to females in the mature part of the cod populations was more than two to one.

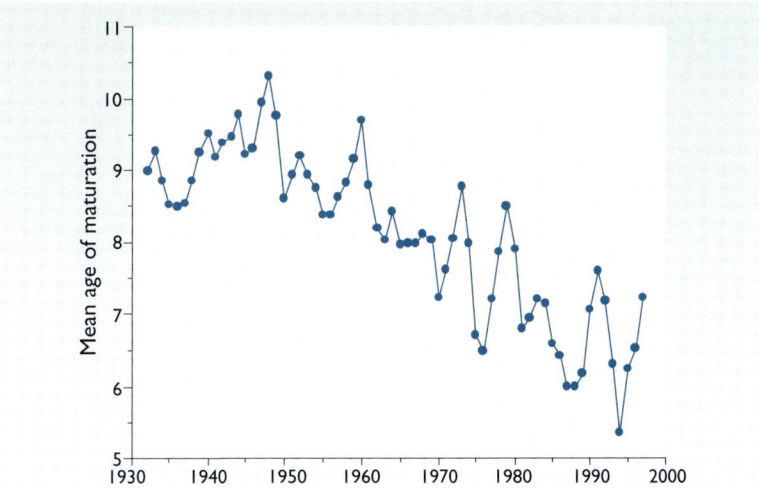

Figure 5.12
Mean age at maturation (Heino et al. 2004).

In Northeast Arctic cod, the average age of maturation has fallen from between nine and eleven years in the 1950s to around six in the 1990s (Jørgensen 1990; Beverton et al. 1994; Heino et al. 2004), as Figure 5.12 shows. Two different mechanisms have been advanced as the cause of the decline. First, density-dependent growth enhancement; i.e. when fishing removes fish from the stock, the competition among the remaining individuals is reduced so that they experience better availability of food, better growth and earlier maturation. In the early to mid-1980s, when the stock, particularly of immature fish, was at an extremely low level (Figure 5.2), the mean length of first-time spawners aged six, seven and eight increased rapidly (Figure 5.10).

Second, a genetic change; i.e. intensive fishing catches late-maturing individuals before they reach maturity so that "the old age maturation genes" are removed from the stock. However, age and size at maturation in fish, including cod, is known from both field and experimental studies to be modified by environmental conditions, in particular by temperature (Godø and Moksness 1987) in addition to the effects of fishing mentioned above, so that the general trend towards warming of the Barents Sea ecosystem since the late 1970s may have contributed to the reduction in age at maturation during the most recent decades.

Since eggs spawned by older cod are more viable than those from young spawners (see next section), the reproduction potential of the stock may have been negatively effected by the development, as shown in Figure 5.12. A better understanding of the causes and mechanisms which have led to that development is urgently needed (Heino et al. 2004).

5.6 Recruitment fluctuations and their causes – environment or parent stock size?

In his summary of the first five years of international cooperation in marine science under the auspices of ICES, Johan Hjort (1908) wrote: "That we can thus from the Arctic ocean and even down to the North Sea trace a connection between the growth and production of fishes and climatic conditions seems to show that the production of fish is subject to such mighty influences that it

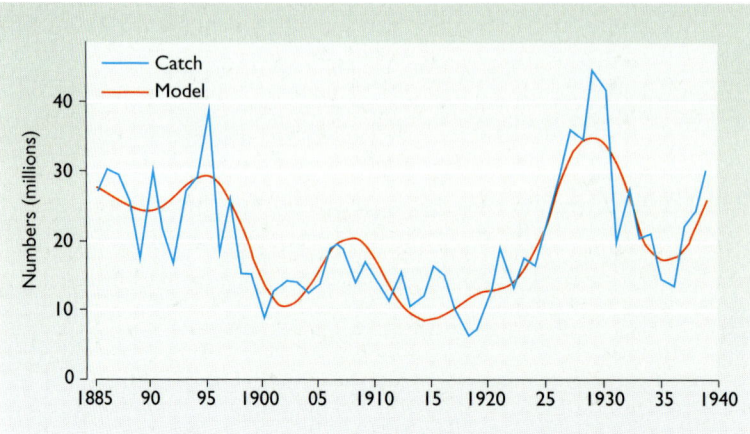

Figure 5.13
The catch of "skrei" in the Lofoten fishery (Ottestad 1942).

may be regarded as independent of the interference of man". The hypothetical conclusion in this statement regarding the effect of fishing (i.e. the interference of mankind) has later been shown to be wrong for exploited species and stocks. However, the general observation, "a connection between the growth and production of fishes and climatic conditions" was elucidated in Hjort's classic work a few years later and became a major topic of study throughout the 20th century. The view that fishing had no significant effects on recruitment was held by most scientists until the 1960s. In his paper "What is over-fishing?", Petersen (1903) distinguished between different effects of fishing, including reducing the mature stock so that too few eggs would be produced to maintain recruitment, and catching undersized fish (too young) with the result that future yields would be reduced. The first possibility was regarded as very unlikely, while the other was a matter of great concern in the North Sea fisheries as early as the turn of the century, and became a major challenge in managing cod fisheries in the North Atlantic throughout the 20th century.

In their invitation to an ICES meeting in 1948 on "Climatic changes in the Arctic in relation to plants and animals", the conveners, Gunnar Rollefsen and Åge Vedel Tåning, wrote: "Thus, it is clear that some of the largest fisheries in the northern hemisphere – e.g. some cod fisheries – during the coming years will be completely dependent on the course of the present mild period in northern waters". The mild period they referred to was the northward penetration of warm saline water masses in the North Atlantic since 1920, which was particularly pronounced during the 1930s and was accompanied by northward displacements of many species of fish as well as greater yields in many fisheries. Obviously, Rollefsen and Tåning were of the opinion that the increased yields of cod during those decades were at least partly the results of increased biomass production due to favourable environmental conditions. Their opinion was based on the results of a number of studies of changes in the North Atlantic ecosystem and their causes, as well as experimental and field work that had demonstrated a positive correlation between fish growth and temperature. Figure 5.13 shows the results of Ottestad (1942), who assumed that cod production, i.e. the yield of the skrei fishery, would be subject to temporal variations similar to those found for the growth of trees in the area, and he obtained a remarkably good agreement between his model results and the actual yield of skrei in Lofoten.

When he discovered the wide variations in year-class strength, Hjort hypothesized that these were due to annual variations in the availability of suitable food for the "very earliest stages of larvae and young fry" (Hjort 1914). Following an idea of G.O. Sars, Rollefsen (1930, 1932) carried out an experiment that led him to suggest that the destruction of eggs and larvae by storm action might also be an important factor in the formation of year-classes. Wiborg (1957), who carried out extensive field studies of newly hatched cod larvae for many years, thought that a long spawning period and a large spawning area favoured the development of abundant year-classes. He also suggested that strong northerly currents during the drift period of eggs and larvae were of importance. Thus, the opinion that recruitment to the stock was exclusively determined by natural causes dominated the discussion, because until the 1960s, there was no sign of any relationship between the amount of spawners (i.e. the yield of the skrei fisheries) and the strength of the year-classes they produced. As to the effects of fishing, scientists were mainly concerned with gross overfishing or wasteful fishing, i.e. where fisheries remove small and young fish to an extent that reduces future yields of large valuable fish. Little or no attention was paid to the problem of recruitment overfishing, i.e. where the fishery reduces the spawning stock so much that recruitment is negatively affected. During the 1960s, however, opinions began to change. Russian studies had indicated a relationship between year-class formation and parent stock size and age structure (see Ponomarenko 1973 for references). Ponomarenko believed that the quantity of eggs spawned as well as their quality was of importance for larval survival and year-class formation, since studies had shown that larvae from repeat spawners (older parents) were more viable than those from first-time spawners. He hypothesized the following relationship between parent stock and year-class strength: "A large spawning stock comprising many age groups will spawn over a larger area and a longer time than a small spawning stock, so that it becomes more likely that some part of the eggs and larvae from a large spawning stock will encounter favourable conditions for survival" (see reference to Wiborg 1957 above). In 1967, the British scientist David Garrod published a spawning stock-recruitment relationship for the stock (Garrod 1967). On the basis of his findings as well as on theoretical considerations, in 1968 ICES expressed concern about the future quantity of spawners unless significant reductions in catches and exploitation rates were obtained; a continuation of the high fishing mortalities experienced throughout the 1960s might leave so few fish left to spawn that the risk of poor recruitment was increased. At the time, this advice was not agreed with by older members of the IMR scientific staff. Because of the large number of eggs that were spawned by each hen cod, they believed that the number of spawners would have to become extremely low before recruitment was affected.

The pessimistic forecast given by ICES in 1968 failed because the very abundant 1963 and 1964 year-classes were underestimated (ICES 1971). When these year-classes became mature in 1969–1972, the spawning stock increased considerably and so did catches of skrei. Moreover, the 1970 year-class was of record high abundance. Thus, stock development in the early 1970s became far more positive than predicted, and this probably persuaded fishermen and managers that the concerns expressed by ICES during the latter half of the 1960s were unjustified.

The physical and biological environment
How many offspring from a spawning reach the size (age) of recruitment to the fishery and how are these numbers reduced as fish become larger (older)?

Ever since the great variations in year-class strength of fish were discovered, this question has been asked. The number of fish that reach the age of three and recruit to the fishable stock varies considerably from year-class to year-class. The 1970 year-class, the largest that has been observed, numbered around 1900 million individuals at that age (Figure 5.2). In contrast, the 1977–1980 year-classes consisted of only 100–200 million individuals at the same age. As mentioned above, it was well known as early as the 1940s, that rising temperatures were favouring recruitment and production in northern cod stocks. At the UN conference on living resources in 1949, Tåning described the effects of the warming of the North Atlantic as follows: "This rise in temperature was especially observed after 1925 when the Arctic water began to retire. Owing to the rise in temperature, immense stretches of banks in northern seas, previously covered with Arctic water, have been made habitable for many species of animals including several species of food fishes normally avoiding Arctic water. By the rise of temperature food fishes, such as the cod, for instance, have obtained an addition to their original area of distribution of thousand and thousand square kilometres, and with this an enormous augmentation of food, i.e. food competition has decreased, enabling an increase in individuals beyond the normal". Sætersdal and Loeng (1987), who studied the relationship between year-class variations in Northeast Arctic cod and temperature in a time series of 80 years, concluded that nearly all high- and medium-abundant year-classes were associated with positive temperature anomalies at the onset of a warm period in the Barents Sea. However, an open question was: When and why do pre-recruits die? Hjort (1914) had hypothesized that the very earliest larvae and young fry died because of a lack of suitable food. Studies of Northeast Arctic cod in the 1970s and 80s supported this hypothesis and led to establishment of the relationships in Figures 5.14 and 5.15. Figure 5.14, which shows the plot of numbers at three years of age (estimated from catch at age data) against temperature at the spawning grounds, clearly shows that recruitment is positively related to temperature. In years with high temperatures, both strong

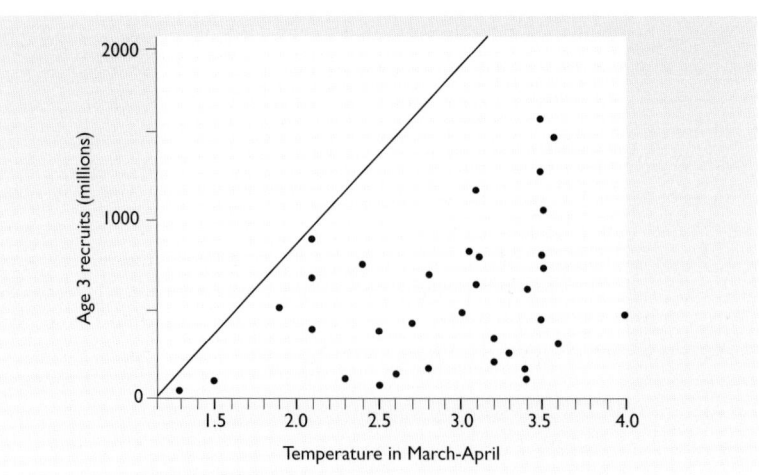

Figure 5.14
Year-class strength at age 3 years against temperature at the spawning grounds (Ellertsen et al. 1989)

Figure 5.15
Time of maximum occurrence of Calanus finmarchicus against temperature at spawning grounds (see Figure 5.14) (Ellertsen et al. 1989)

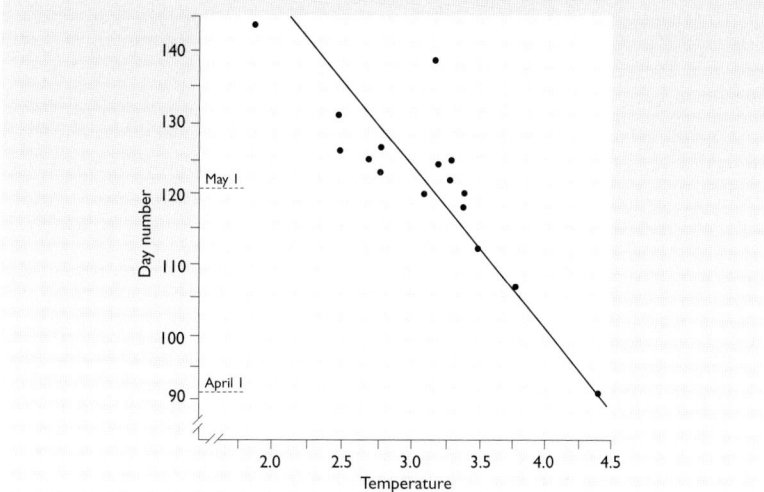

and weak year-classes are produced while strong year-classes are absent when the temperature in the area is low. Figure 5.15 show the time of occurrence of fry of *Calanus finmarchicus*, the main prey item for cod larvae, and suggest a possible contributor for the relationship in Figure 5.14. The parent stock of Calanus overwinters in deep waters in the North Atlantic and the Norwegian Sea. They ascend to surface waters to spawn during late winter and early spring, when the spring bloom occurs. Their spawning time is closely related to temperature. At low temperatures they spawn so late in spring that most cod larvae, which hatch in April, do not find food (Ellertsen *et al.* 1989; Solemdal 1997; Sundby 2000).

A significant contribution to our understanding of the survival of fish larvae is the knowledge of how turbulence affects contact rates between larvae and their prey (Rotschild and Osborn 1988; Sundby and Fossum 1990). Up to a certain level, turbulence favours larvae feeding by increasing these contact rates.

Recruitment in cod is probably one of the most thoroughly investigated subjects in the history of marine science. As long as 30 years ago, there was evidence that the recruitment of cod in the North Sea was decreasing with increasing temperature (Dickson *et al.* 1974) while there were indications that higher temperatures favoured recruitment in the more northerly cod stocks.

Studies carried out during the 1990s, have clarified this apparent contradiction in the relationship between temperature and recruitment for southern and northern cod stocks as Figure 5.16 shows. While southern cod stocks (North Sea, Irish Sea), which inhabit warmer waters, have less recruitment in warm years than in cold years, the opposite is true for northern stocks situated in cold waters. These stocks (Northeast Arctic, West Greenland, Island) have better recruitment in warm years than in cold (Ottersen 1996). In his comprehensive review of the subject, Sundby (2000) proposed that this difference is caused by changes in the transport of the copepod Calanus from their core areas in the Atlantic and Norwegian Sea to the areas where the various cod stocks have their habitat; i.e. that the apparent temperature–recruitment relationship in Figure 5.14 is a proxy for a food abundance–recruitment relationship due to wide variations in transport of suitable food for cod larvae.

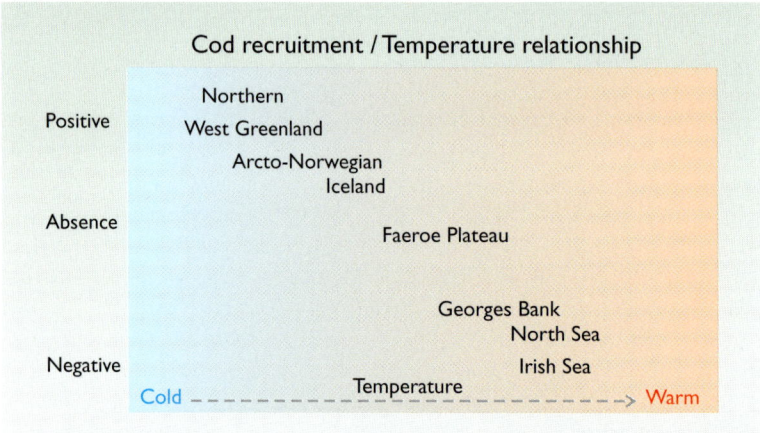

Figure 5.16
The nature of the relationship between cod recruitment and temperature changes for cod stocks in the Atlantic (Ottersen 1996; Planque and Fox 1998).

Another important source of cod mortality during the first two or three years of its life is predation. For the fry, the predation pressure increases in autumn during the first year of life when they migrate deeper in the water and become available to larger fish, predominantly older cod. In periods when the main stock of prey fish (capelin) is small, predation on cod juveniles, including cannibalism, increases. Thus, in 1986–1988, when capelin were scarce, cannibalism reduced the 1985- and 1986 year-classes which were very abundant as 0-group, to far below average abundance as three-year-olds. As early as the mid-1970s, Ponomarenko and Ponomarenko (1975) anticipated these circumstances. On the basis of estimates of the food demands of the cod and haddock stocks and the production capacity of the capelin stock, they concluded that if the two gadoid stocks were to recover, the capelin stock and fishery would decline and the cod would have to change to other food. In the event, the stock of capelin was heavily reduced from 1984 to 1986, and fishing was banned from 1987 to 1989; the cod changed to other sources of food than capelin, including greater dependence on cannibalism (Hamre 1988, 1991; Mehl 1989, 1991). IMR had initiated studies on interactions between cod and other species in the Barents Sea in the early 1980s. The collapse of the capelin stock in the mid-1980s and its consequences for the recruitment and growth, particularly in cod, led to an increase in effort and a close cooperation with PINRO on the subject. Within a few years, a cod/capelin interrelationship was established, so that since the early 1990s, the cod's need for capelin has been taken into account when capelin TACs are recommended by ICES. Moreover, annual estimates of the number of cod (by age) eaten by cod are available back to 1984. These estimates are based on data from the comprehensive stomach sampling project in the multispecies stock modelling programme (MULTSPEC) at IMR and PINRO, and they are used in the annual stock assessments conducted by ICES (see Bogstad and Tjelmeland 1992a, b; Tjelmeland and Bogstad 1998).

The parent stock – Numbers, age composition and condition of spawners
Until the 1960s, it was, as stated previously, assumed that since each hen cod spawns several million eggs, the number of parents was of little or none importance for the abundance of a year-class (see above). It was thought that lack of fish would make commercial fisheries unprofitable at stock levels well

above those at which recruitment was affected. Before the 1950s, neither the available observations nor the analytical tools were suitable for analysing stock-recruitment relationships and determining possible recruitment overfishing. The development of the theory of fish population dynamics and its implementation using catch at age data changed this situation. In the 1960s, it became possible to routinely generate life-history tables for each year-class, making time series with annual estimates of parents and offspring (i.e. recruits to the fishery at age three), and thus to quantify the stock-recruitment relationship. For cod the first known attempts at such quantification were made by British scientists (Garrod 1967; Garrod and Jones 1974). These studies showed that throughout the 1950s and 60s, the fisheries had reduced the spawning stock to such low levels that recruitment was probably affected, and they estimated an exploitation rate which should not be exceeded in order to keep the number of spawners at levels that would maintain recruitment. Their findings were largely supported by subsequent Norwegian and Russian studies (Jakobsen 1992 and 1993; Serebryakov 1991), which indicated that in order to maintain sufficient levels of recruitment a spawning biomass of 500 000 tonnes or more is needed. Jakobsen (1993) also concluded that a lower level of exploitation than that corresponding to a fishing mortality of 0.46 should be aimed at (Garrod and Jones (1974) had suggested a level of 0.43). Figure 5.17 shows the spawning stock–recruitment relationship established by Jakobsen (1996).

Solemdal (1997) summarized and reviewed results of recruitment studies in cod for 1970–1996. The following is a brief summary of the main findings regarding the importance of the spawners for year-class formation: It has been shown that second- and third-time spawners and females in good condition generally spawn larger and better-nourished eggs than first-time spawners and fish in poor condition. The larvae from "old" spawners may therefore feed longer on the yolk sac and their chances of survival will thus increase. Furthermore, eggs from repeat spawners show greater variation in buoyancy and will therefore be more dispersed vertically and horizontally, thereby ensuring that hatching takes place over a larger area in the sea and thus increasing the possibility of good feeding and better survival conditions for some of the larvae. A large spawning stock consisting of several age groups – both first-time spawners and those which have spawned several times – will spawn over a larger area and a longer period of time than a small and young spawning stock. In consequence, the possibility that cod larvae will experience good conditions for survival increases with the

Figure 5.17
Northeast Arctic cod. Spawning stock biomass and recruitment at age 3 (Jakobsen 1996).

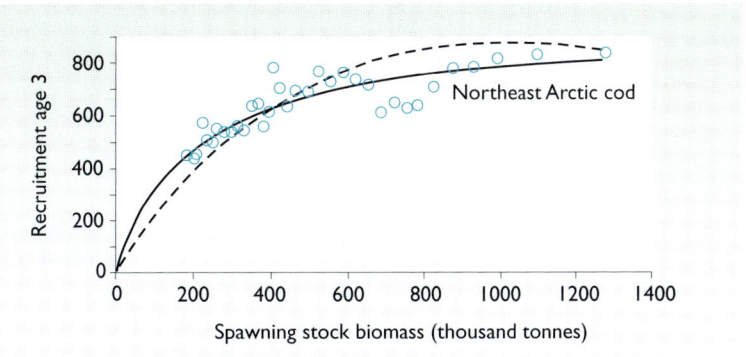

size of the spawning stock and its average age and condition (see the reference to Ponomarenko 1973 above).

At an ICES symposium on cod and climate in 1992, Ottersen *et al.* (1994) argued that the change in stock structure caused by intense exploitation over 50 years had made the stock more sensitive to environmental fluctuations. This view was supported by Daan (1994), who in a review of trends in several cod stocks wrote: "When the level of exploitation is increased, the number of year-classes contributing to the spawning stock decreases correspondingly, and at high fishing mortalities, short-term variations in year-class strength are transferred directly into short-term variations in spawning stock. Under these conditions, adverse climate events of rather shorter duration could easily result in stock collapses. In other words, exploitation leads axiomatically to a reduced resilience of a stock against effects of climate".

It seems appropriate to end this short review of recruitment fluctuations and their causes by quoting some sentences from a study by Ottersen and Sundby (1995): "Of the environmental parameters considered in this study, temperature is the one most significantly related to cod recruitment success. However, it should be emphasized that the results of this paper indicate that the spawning stock biomass is a factor nearly as important to the formation of year-class strength. These results show the fruitlessness of the old simplistic discussion of whether recruitment variability is environmentally driven or driven by the spawning stock biomass. Both environmental and spawning stock aspects should be considered".

Efforts to establish levels and reference limits of spawning stock biomass and exploitation rates in order to maintain recruitment and sustainable catches were greatly increased for most stocks by ICES during the 1990s. For Northeast Arctic cod these efforts resulted in precautionary and critical levels of spawning stock biomass of 460 000 and 220 000 tonnes, respectively, and in fishing mortality rates of 0.4 (precautionary) and 0.74 (critical). These levels were adopted by the Norwegian-Russian Mixed Fishery Commission (see section on management). However, it should be noted that the size and age structure of the female component of the spawning stock represents a more relevant measure of the reproductive and recruitment potential of Northeast Arctic cod than the traditional spawning stock biomass which includes both females and males and takes no account of its age composition (Marshall *et al.* 1998; Ajiad *et al.* 1999).

5.7 Exploitation, stock monitoring, advice and management
1900–1940

Early in the 20th century, the rate of exploitation (i.e. the fishing mortality) was not regarded as important for stock development and future yields. Cod tagging experiments in 1913 indicated that the skrei fisheries produced a mortality of about 20–25 percent of adult fish; "… every fourth or fifth fish was recaptured in the Lofoten fishery in 1913" (Hjort 1914). Hjort discussed the effect of such a mortality rate on the average longevity of the fish and the reduction in numbers by age of a rich year-class, and concluded that: "… a rich new stock of seven-year-old fish would after four or five years intensive fishing be considerably reduced in numbers". He also asked whether this could have an effect on recruitment, i.e. would the spawning stock be reduced to a level at which recruitment would decrease. However, compared with the great variations in year-class strength, he regarded the effects of the fishery as negligible: "The renewal of the stock can thus scarcely be dependent upon so regular and constant a factor as the fishing; it must depend upon highly variable natural conditions".

The new knowledge that a strong year-class made up the bulk of the annual catches for several years in succession was utilized to forecast the availability and size of fish in the skrei fisheries. A sampling programme for monitoring the age and size composition of catches was introduced in the Finnmark (immature fish) and Lofoten (mature fish) fisheries, yielding a nearly unbroken time series back to 1913 (Hylen 2002). Figure 5.18 shows the yearly length distributions of catches in the two fisheries for 1913–1935 in terms of deviations from the mean value for the whole period (Sund 1936). Sund identified four outstanding year-classes (1904, 1912, 1917 and 1919) from these data; the fish entering the Finnmark fishery at a length of 40 cm (aged 4–6 years) and then entering the spawning cod fishery on the Lofoten grounds two or three years later. However, uncertainties regarding the age determinations, which at that time were based on scale readings, and lack of observations from the ocean regions of the Barents Sea, weakened the reliability of the predictions and limited their value for forecasting the skrei fisheries (Sund 1936; Rollefsen 1938). In addition, the shift in exploitation pattern towards younger and smaller fish caused by the rapidly increasing trawl fisheries in the area during the 1920s made comparisons over the entire period difficult.

Throughout the 1930s, Norwegian scientists became increasingly concerned about the possible effects of the growing international trawl fisheries in the Barents Sea area. Their concerns were twofold. First, they regarded trawl fishing as wasteful because of the large amounts of small cod and haddock of no commercial value that were killed. This called for the use of size limits and larger mesh sizes, a subject that was then discussed in ICES for the North Sea fisheries (see Chapter 2). Secondly, they feared that the exploitation had reached levels that could reduce the availability of fish and thus yields in the Norwegian coastal fisheries. Thor Iversen, who carried out surveys using a commercial ground trawl in the area in 1932 and 1933 and found that large quantities of small sized no marketable fish had to be discarded at sea, wrote: "It is a question whether this population can stand such excessive exploitation as this powerful depletion of fry of food fish signifies" (Iversen 1933).

In a correspondence with the prime minister in 1934, Johan Hjort, by then a professor at the University of Oslo, expressed support for the development of Norwegian trawl fisheries in the area. Hjort had always been in favour of modernizing Norwegian fisheries and recommended trawling, always provided measures could be taken to avoid wasteful fishing. Oscar Sund, on the other hand, who was the head of cod investigations at the Directorate of Fisheries, was extremely sceptical. He wrote that the trawl fisheries during the previous decade (1924–1933) had probably tripled the exploitation rate of cod and haddock in the area. Although there were no signs of decreasing yields in the Norwegian coastal fisheries, he thought that comprehensive studies of age distributions of the catches should be carried out before Norwegian trawl fisheries were supported. By analysing frequency distributions of annual catch at age based on otolith readings, Rollefsen (1937) found a substantial increase in the total mortality of adult fish during the 1930s, from about 40 percent in 1932–1934 to about 60

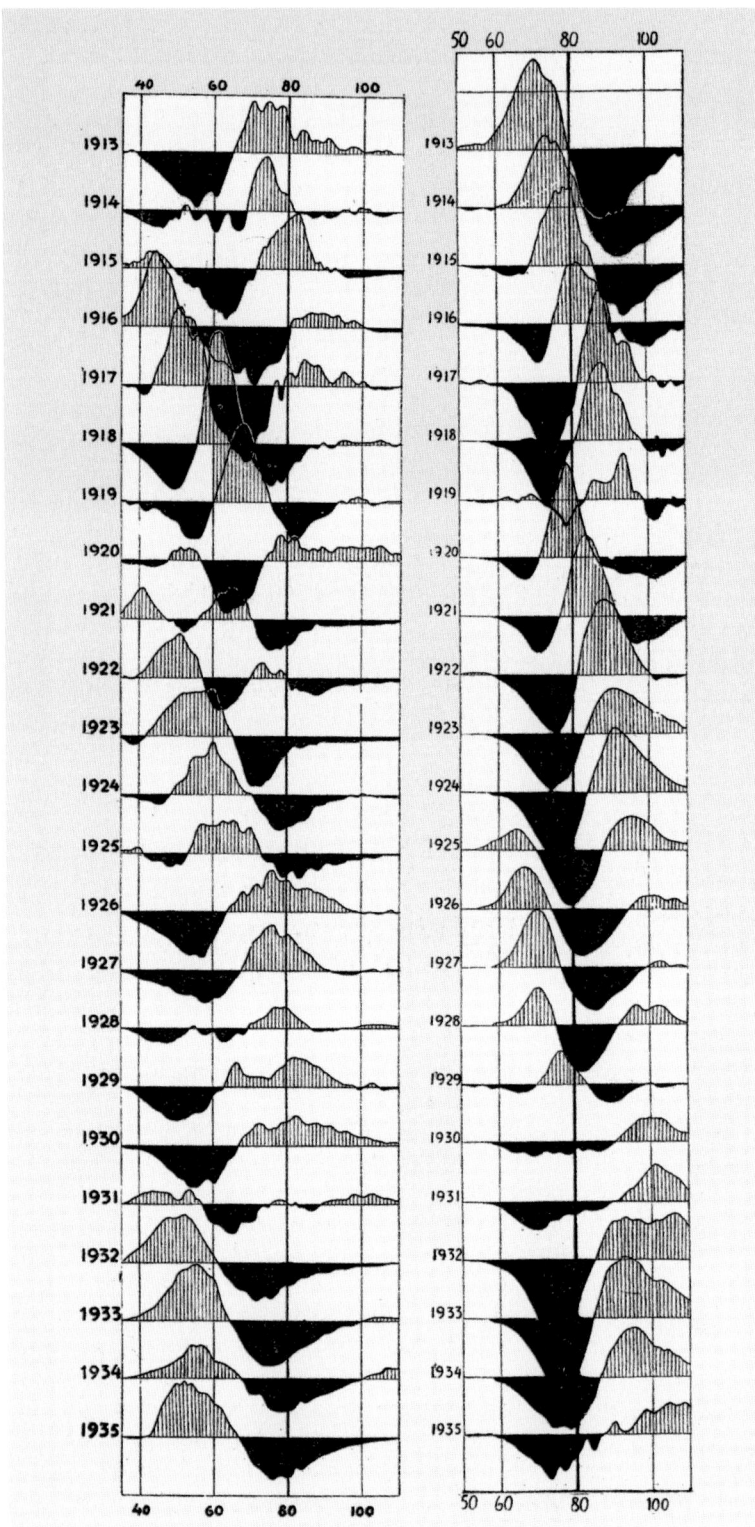

Figure 5.18
Monitoring year-class strength from length distribution of catches (Sund 1936).

percent in 1935–1937, and he attributed both the increased mortality and an observed reduction in the mean age of the spawning cod (skrei) to the increased yields in the international trawl fisheries in the 1930s. He wrote: "Apart from a possible reduction of the Norwegian yield of skrei by foreign trawling, we must also regard this new fishery as an additional levy on the stock, and it does not appear unreasonable to assume that this extra toll has been a contributory cause of the reduction of the mean age of the skrei" (Rollefsen 1938).

In 1936, the "Over-fishing Conference" on mesh size regulations and size limits for fish was held in London. Participants from Norway included Johan Hjort and Thor Iversen. Obviously, there was a recognition at the conference that due to their slow growth and late maturation, cod and haddock in northern waters needed extra protection, because in the draft conclusions a minimum mesh size of 105 mm was recommended in trawls for the area north of 66°N and east of 0°E (Barents Sea), as compared with 70 mm in more southerly areas (North Sea). However, in the International Convention of 1937 for the regulation of meshes in fishing nets and size limits of fish, the mesh size was agreed at 80 mm. This agreement never formally came into effect. The USSR did not attend the meeting in 1937, and by the outbreak of the Second World War in 1939, several countries had not ratified the agreement (Engesæter 2002).

1940–1977

During most of World War II the Barents Sea was a war zone and little or no offshore fishing took place. The total annual catch of cod declined from about 800 000 tonnes in the immediate pre-war years to around 200 000 tonnes in 1943–1944, exclusively taken in the Norwegian coastal fisheries. The reduced fishing mortality allowed the stock to recover, and after the war, when fishing effort increased, catches and exploitation rates reached pre-war levels within few years (Figure 5.2).

A second international conference on over-fishing, held in London in 1946, proposed a convention for the Northeast Atlantic, which in addition to regulating mesh sizes and size limits, included closed areas and seasons, as well as restrictions on fleet capacity. In an article written just after that conference, Gunnar Rollefsen stated that the stock of cod in the Barents Sea area during the past 20 years had been very large in comparison with previous decades. He expected that the yields in the Norwegian traditional fisheries would decline in the years to come due to increased foreign trawling in the area, particularly so if the production in the stock was reduced to the level of before 1920. He also pointed to the possibility that catch limits would be introduced and that fishing nations would be allocated quotas according to their historical catches (Rollefsen 1946). However, quotas, or fleet capacity allocations, were still too radical tools for international management, and the 1946 Convention for the Northeast Atlantic fisheries and its "Permanent Commission" – established in 1953 – was severely limited by lack of power. It could have been a useful tool for the pre-war fisheries, but the post-war period became very dynamic, with greatly expanding fisheries and new fishing nations emerging (Sætersdal 1992). By 1955, total annual landings of Northeast Arctic cod had reached 1.3 million tonnes due to a substantial increase in fishing effort, in particular in the international offshore trawl fisheries. To adapt to this development, a new convention with wider powers was agreed in 1962 by the European fishing nations, with NEAFC (the Northeast Atlantic Fisheries Commission) replacing the Permanent Commission (Engesæter 2002).

In 1957, IMR carried out a study of Norwegian data on Northeast Arctic cod, which indicated an unusually low yield in the skrei fisheries during the 1950s compared with the yields of immature fish of the same year-classes (Sætersdal and Hylen 1964). Their main conclusions were that: "…the skrei population has apparently been less abundant compared to the population of young cod, than it was before the war. This is the type of effect we must expect to find if the reduction of the stock of skrei is not largely a result of natural fluctuations, but has been caused by an increase in the total exploitation of the Arctic cod. The number of old and large fish is reduced relatively to that of the younger and smaller fish". They also found that the mean age of spawning cod had fallen during the same period as the reduction in abundance, and interpreted this as an effect of increased exploitation. The study was submitted to the 6th meeting of the "Permanent Commission", which asked the ICES Liaison Committee to encourage an investigation of all available international data on the dynamics of the Northeast Arctic cod stock and associated fisheries. The outcome was the establishment of the ICES Arctic Fisheries Working Group, which had its first meeting in Bergen in 1959 (see Hylen 2002 for more details). This was a milestone in monitoring the stock. Participants from the UK, Germany, Norway and the USSR, the major Northeast Arctic cod fishing nations, supplied relevant data on cod (and haddock) for the period 1930–1958, so that data from all fleets and fisheries for a long period could be jointly analysed. The results of the analyses largely supported the findings of Sætersdal and Hylen.

The Arctic Fisheries Working Group has met every year since 1959, except 1961, sometimes twice a year. Its main responsibility has been, and still is, to produce annual assessments of the state of stocks and proposals for the advice ICES should offer regarding the regulations of the fisheries. In the early 1960s, the Group was mainly concerned with the large amounts of small fish, cod in particular, which were caught and largely discarded at sea, and recommended increasing the minimum size of catchable fish as well as larger mesh sizes, and considerable effort was put into mesh selection experiments in order to establish selection factors for the various types of net material in use. However, it was soon realized that the increasing fishing effort during the 1950s and 60s had brought the exploitation rate of cod to levels that the stock could not sustain, and the 1965 report from the Group (ICES 1965) included a presentation of measures for limiting catches (closed areas, increased mesh size, reduced fishing effort and catch quotas). The Norwegian Director of Fisheries, Klaus Sunnanaa, expressed satisfaction with the advice given by ICES, but he was frustrated by the inability of NEAFC to implement it in practical fisheries management, as he feared that Michael Graham's "Great Law of Fishing" would apply (see Chapter 2). At the 9th Nordic Fishing Conference in Reykjavik in June 1964, he gave a speech with the following main message: Since 1946, the only result of international cooperation on management of the stock of Northeast Arctic cod was a minor increase in mesh size in trawls, a measure which to a large extent was rendered ineffective by technical arrangements of the gear. The existing fishery regulations, i.e. minimum mesh size and minimum legal size of fish, were not sufficient to prevent depletion of stocks and yields with possible serious future consequences for coastal states and districts. Work aimed at total catch limits for some stocks and areas should therefore start immediately. If catch limits could not become effective within a short time, the coastal states would probably initiate efforts to extend their jurisdiction to include the entire continental shelf in order to improve fisheries management.

From 1968 on, ICES consistently expressed concern at the future size of the spawning stock, considering that at low spawning stock levels, the risk of poor recruitment was increased. The increase in the spawning stock in the early 1970s caused by the abundant 1963 and 1964 year-classes, and probably also the scepticism of some scientists (see above) regarding the parent stock-recruitment relationship, probably persuaded managers that the concerns of ICES were unjustified. Exploitation rates remained high, particularly for young and immature fish aged 3–6 years. In a comprehensive analysis presented to ICES in 1975, Hylen and Rørvik (1975) demonstrated that total annual yields as well as spawning stock biomass would increase if catches of young cod were reduced by increasing trawl mesh size and thus size and age at capture. Although their results were supported by ICES, no measures were introduced to increase the survival rates of young cod.

The fact that the largest year-class ever recorded, the 1970 year-class, recruited to the fisheries in 1973–1974 and raised catches for some years, probably helped to weaken the effect of the warnings of the scientists. However, the fate of the 1970 year-class and the development of the stock during the 1970s and early 1980s were even worse than the scientists had predicted (Figure 5.2), mainly because no effective management measures were introduced prior to the establishment of the national economic zones (NEZ) in 1977. The first catch quotas (TAC) for cod, which were introduced under the NEAFC regime in 1975, were 810 000 tonnes and were far too high. A similar TAC was agreed for 1976. These TACs were at levels recommended by ICES. However, the stock estimates and predictions that formed the basis of the advice were too "optimistic", so that the decline in stock size and catches towards the end of the 1970s was much greater than expected. Thus, the situation in 1977, when Norway and Russia (USSR) were given responsibility for fisheries management in the area, can be summarized as follows: landings of cod were about to decline due to a rapidly declining stock, and the rates of exploitation were far above any sustainable level (Sætersdal 1989).

1977–2000

An analysis performed by Ulltang (1987) showed that both yields and stock size could have been maintained at higher levels in 1977–1985 if adequate management measures had been introduced and enforced throughout the 1970s. However, the cod belonging to the relatively abundant year-classes in 1969–1975, which could have maintained the stock in that period, were largely caught as young and immature fish, only 3–5 years of age.

What had gone wrong? Since the early 1960s, a well established and fully recognized science-based management system through ICES and NEAFC had been available for the area. And unlike the situation in other fisheries, including those for Norwegian spring-spawning herring, ICES had recommended reductions in the exploitation rate of cod in most of these years. However, apart from a slight increase in mesh size in the trawl fisheries, the parties concerned had been unable to act in unison for a common long-term benefit against short-term individual interests. The emerging Law of the Sea Convention, with its drastic shift of interests to the coastal states, probably contributed to the poor results of the work of NEAFC during the final years of its existence (Sætersdal 1993).

We may also ask to what extent the "pessimistic" forecasts given by the scientists at the end of the 1960s were acknowledged by the authorities. The failure of ICES to predict the substantial increase in stock and catches of cod in 1968–1972, caused by the very abundant 1963 and 1964 year-classes, probably weakened the confidence of fishermen and politicians in its advice. The main cause of the failure was the lack of reliable information on the abundance of year-classes before they entered the fishery at 3–5 years of age. Stock estimates and predictions were exclusively based on data from the commercial fisheries, data which yielded little or no information on year-class strength before the fish had appeared in the fisheries for some time. In order to improve predictions, fisheries-independent indices of abundance of young fish were needed, and in the early 1970s, IMR introduced surveys in the area that aimed to provide acoustic abundance estimates of the younger age groups of cod and haddock (Hylen and Smedstad 1972). For some years, exploratory surveys were conducted in different seasons to evaluate the time of the year when the spatial distribution of fish was favourable for acoustic surveying, and since 1976, acoustic surveys have been carried out annually in the area in January–March, with the objective of obtaining estimates of numbers at age of cod and haddock. Survey effort was increased substantially in 1981 when the acoustic survey was combined with a bottom trawl survey designed to provide "swept-area" estimates of abundance. The primary reason for the increase in survey effort was the difficulties experienced in the ICES' stock assessments of cod in 1977–1980. During those years there were large discrepancies between the information obtained from commercial catch data and the acoustic survey results regarding stock status and development, with the survey results indicating substantially higher mortalities of fish (and thus exploitation rates) than did the analysis of the commercial catch data. Because of these inconsistencies, the Advisory Committee on Fisheries Management (ACFM) in ICES made the following recommendation to the Arctic Fisheries Working Group (ICES 1980): "In estimating fishing mortalities and stock sizes in 1979 and 1980, more weight should be given to reliable survey results, particularly if two independent surveys are in reasonable agreement, than to fishery-dependent data". Another reason for increasing the survey activity in 1981 was the necessity to closely monitor the strength of the 1976–1980 year-classes, all of which had been observed to be of low or very low abundance in the 0-group surveys. This suggested that the fishable stock in 1980–1985 might become extremely low unless severe reductions in catch quotas were introduced.

The combined acoustic-bottom trawl survey in the Barents Sea in winter has been performed annually since 1981, yielding two parallel time series of estimates or indices of abundance at age (Figure 5.19). These estimates have become the most important fishery-independent data used to assess cod stocks. Jakobsen *et al.* (1997) reviewed and summarized the methodology and data processing methods employed as well as the use of survey information for various purposes including stock assessment, environmental monitoring and studies of species interaction. They concluded that the data provided by the surveys had been of invaluable importance for monitoring and assessing the two stocks. Until 1984,

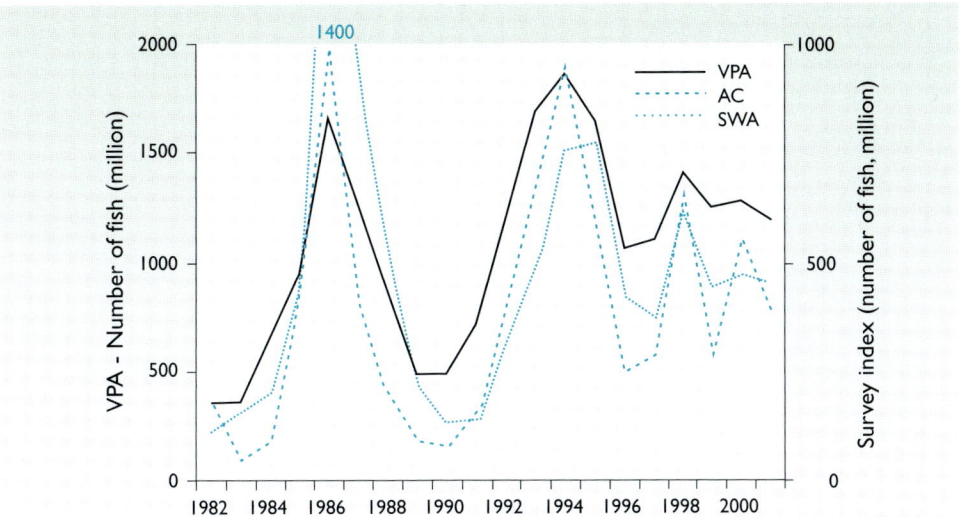

Figure 5.19
Northeast Arctic cod. Survey indices of abundance of 3–5 year old fish compared with results from the catch at age analyses (VPA) at ICES. AC: Acoustic indices, SWA: Swept area indices (Data from ICES, 2004).

the survey results were used mainly for comparison with the results from the traditional assessment method; a procedure in which commercial catch statistics and catch per unit of effort data were used to estimate fishing mortalities and stock numbers. Since 1984, both sets of survey indices (acoustic and swept area) have been used directly in the assessment together with the other data. Hylen *et al.* (1986), who discussed the various sources of error associated with both types of survey index, found that "inefficient sampling of the acoustic recordings is by far the largest source of error in the acoustic index". Their results stimulated investigations into the catching efficiency of demersal sampling trawls in Norway (see Chapter 8), investigations which led to substantial improvements in gear and methodology (Engås 1994) and the resultant abundance indices (Aglen and Nakken 1997); improvements which necessitated revisions of earlier established indices in order to maintain comparability in the time series.

Since Norway and Russia were given the responsibility for managing the fisheries, the main steps in the advice and decision making process have been the following (Nakken *et al.* 1996; Aglen *et al.* 2004):

1. Institutes of marine research in the two countries conduct surveys every year, observing abundance and age composition of the stock. The institutes also collect data on the age composition of the commercial catches. All these data are annually assessed by the ICES' Arctic Fisheries Working Group. The ICES' Advisory Committee on Fisheries Management (ACFM) reviews the Working Group report and recommends annual TACs and TAC options, including tables that demonstrate the consequences of the various options on further development of total and spawning stock biomass.
2. The advice or the options presented by ICES are analysed by fisheries management bodies in Norway (and Russia) in order to determine the level of fishing and TAC which seems most beneficial to Norway (and Russia). Norway and Russia, in the meeting of the Mixed Fishery Commission, finally agree on TACs and any other management measures. The TAC is divided equally between Norway and Russia, and allocations are made to third nations (EU, Faroe Islands, Greenland and Iceland).

The Norwegian quota is split between different vessel groups and individual vessels. Approximately 30 percent of the Norwegian quota is allocated to trawlers, while the remainder is allocated to vessels using conventional gears (gillnets, longlines, handlines, Danish seine, etc).

The management of the fishery includes various measures to protect small cod from being caught as well as TAC, and a brief review of each of these points is given below.

Protection of small fish
Since the international trawl fishery developed in the area in the 1920s, a main objective for the Norwegian fisheries authorities has been to limit the catch of young and small cod in order to ensure the continued availability of large, commercially valuable fish and yields in the coastal fisheries. Their view was supported by yield per recruit calculations, which indicated that the highest yield from the stock would be obtained if the cod were left unfished until it had reached a size of about 50 cm (an age of approximately five years) corresponding to a mesh size in trawls of 150 mm or more. During the 1920s and 30s, mesh sizes in use were probably 60 mm or less, and vast quantities of fish in the length range 20–40 cm were caught and discarded at sea (see reference to Thor Iversen). During the first decades after World War II, substantial amounts of small cod were also sometimes discarded, particularly in years when abundant year-classes had a mean length at – or just below – the commercial size (Hylen 1966 and 1967; Hylen and Smedstad 1974). Dingsør (2001), who reviewed and analysed the available information on discards for the period 1946–1998, found that before 1970, stock numbers at three years of age for some year-classes ought to be raised by 25–40 percent when discards were taken into account. Since 1977, discarding of cod has been prohibited. Even so, in the late 1980s, considerable amounts of young cod were discarded by Norwegian trawlers (Hylen 1987), and in 1998, large proportions of the trawl catches in the Bear Island–Svalbard area were discarded because the fish were too small (ICES 1999).

The legal minimum mesh size in trawl gear in the area has been increased several times after World War II, from 80 mm in the 1950s to 135 mm and 125 mm in 1982 in the Norwegian (NEZ) and Russian zone (REZ) respectively, and since 1982, minimum legal fish lengths have thus been 47 cm and 42 cm. The reason behind these differences in measures is the size distribution of fish in the cod fisheries in the two areas. The eastern Barents Sea (REZ) is the main young fish area for the stock where fisheries traditionally are based on young and immature fish, while most of the cod fished in NEZ are larger mature fish. In 1997, sorting grids with a minimum spacing of 55 mm were made mandatory in trawl fisheries for cod in the whole area (see Chapter 8).

In addition to mesh size regulations, an area closure system was established in the early 1980s; areas where the amount of undersized fish in the catches exceeds a certain limit are closed and reopened on the basis of information from monitoring surveys. The very first area closure for cod fishery management purposes in the Barents Sea took place in spring 1979. Then Norway closed a large area (the North Cape bank) for fishing for several weeks in order to protect the cod of the 1975 year-class which then was about 40 cm long. This was an unexpected and rather dramatic action for the international fishing fleet which were operating in the area (Figure 5.20).

Figure 5.20
Title in Fishing News,
27 April 1979.

KICKED OUT OF NORWAY

THE DISTANT water fleet has suffered another major blow as a result of Norway ordering trawlers out of a large area of the north-east Arctic grounds.

Shrimp fisheries with small-mesh bottom trawls developed rapidly in the Barents Sea region, from landings amounting to some few thousand tonnes in the mid-1970s to 126 000 tonnes in 1984. Shrimp trawling may catch large quantities of small fish which are discarded at sea and thus introduce a substantial mortality in various species and stocks. During the 1970s, IMR performed annual surveys on shrimp fishing grounds in order to recommend measures for the protection of species of commercial value, including cod. In 1982, a special monitoring service unit was established under the Directorate of Fisheries, with responsibility to observe the fishing fields and advise about area closures in the cod and shrimp fisheries. Since then the shrimp fisheries have been managed using a closed area system in order to limit the catch of small cod and haddock. In more recent years, redfish and Greenland halibut have also been included in the system; areas are closed and reopened for shrimp trawling according to the by-catch of these species in the catches. Sorting grids for the purpose of avoiding catches of fish (cod, haddock, redfish and Greenland halibut) have been mandatory in the shrimp fisheries since 1992.

TACs (Total allowable catches)
Nakken (1998) reviewed developments regarding advice and management for TACs during the 1980s and 90s of several stocks, including Northeast Arctic cod. He found that most of the annual assessments for cod had underestimated the exploitation rate and overestimated stock numbers and biomass. Hence the annually recommended TACs, which were based on the assessments, had been too "optimistic" or too high. There had also been a tendency for the final TACs to be higher than advised, and finally, the actual catch often exceeded the agreed TAC. The following paragraphs discuss each of these points in more detail.

In the late 1970s and early 1980s, catch per unit of effort in the commercial fisheries was used as a measure of stock abundance. Between 1978 and 1982, the Barents Sea was extremely cold, and the cod concentrated in the south-western part of the ocean where the main fisheries also took place. Fish densities in the area and catch per unit of effort were thus maintained, although the stock declined rapidly (Figure 5.21).

In the mid-1980s, ICES predicted, on the basis of high abundance indices of 0–2 year-old fish from survey observations, a substantial increase in stock size and catches by the end of the 1980s. The predictions failed completely because of interrelationships with other stocks, which the scientists did not

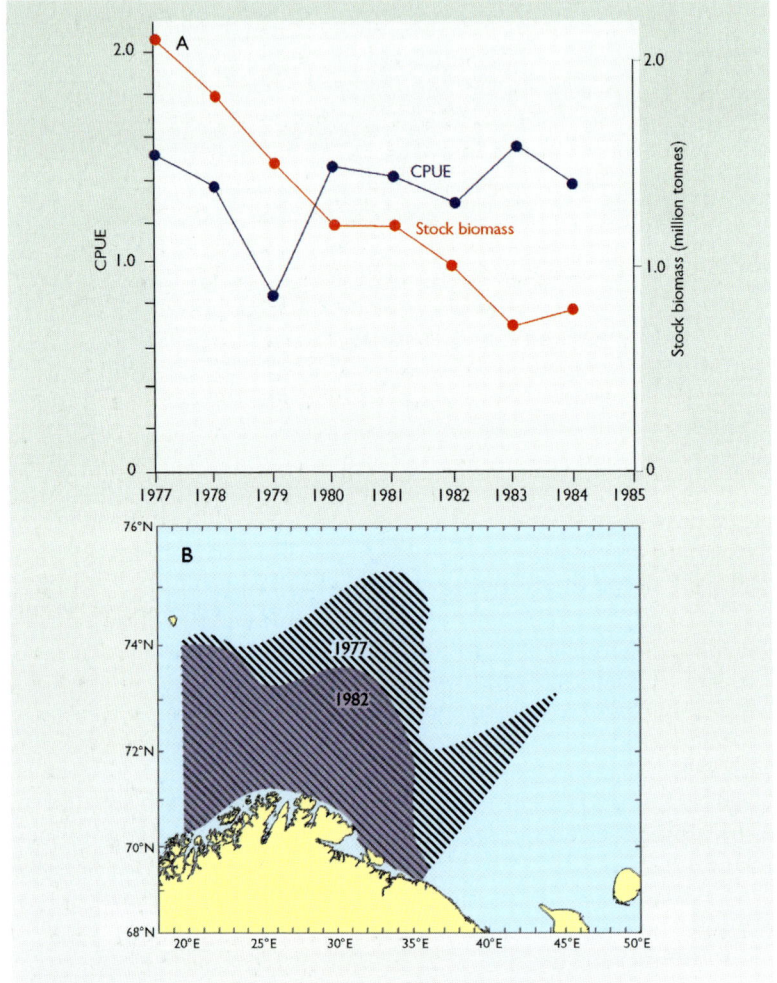

Figure 5.21
Northeast Arctic cod. Upper: The development of catch per unit of effort (CPUE) for Norwegian trawlers compared with the development of stock biomass during a period of intense cooling of the Barents Sea. Lower: The extension of the distribution of cod in February; 1977 (warm) and 1982 (cold) (Nakken 2002).

foresee (see 5.5 Growth and maturation). Recommended TACs for these years (1985–1988) were far too high, and the stock was fished down to a record low level. Ulltang (1996) discussed the uncertainties related to the estimation and prediction of stock size. His main conclusion was that if all available biological and environmental knowledge had been systematized and used, then failures in predicting stock size like the ones experienced in these years might have been avoided, at least to some extent.

Until 1989, the annual TACs agreed on by Norway and USSR included only the trawl fisheries. Vessels using traditional or conventional gears (handline, longline, gillnet, Danish seine) were not restricted by annual quotas of Northeast Arctic cod. Since the main part (60–80 percent) of the Norwegian catch were taken by conventional gears, the actual Norwegian catch largely exceeded the agreed Norwegian "TAC" in most of the 1980s. At the end of the decade, when it was evident that the stock was in a very poor condition, quotas were drastically reduced and all vessel groups included in TAC regulations.

Low TACs in 1990–1991, which resulted in the lowest fishing mortalities experienced since the Second World War, protected the 1983 year-class so that it brought a substantial increase in spawning stock biomass in the early 1990s, and a series of relatively abundant year-classes (1989–1991) helped the stock to recover during the first half of the 1990s (Figure 5.2).

In 1996, the scientists again found large discrepancies between the two main sources of data, abundance surveys and commercial catches, regarding the size and development of the stock, with the survey results indicating higher mortalities and lower stock numbers than did the catch data. During the assessment, more weight was given to the catch data, and this resulted in a major underestimation of mortalities in 1995–1996 and a prediction of stock sizes and TACs for the years to come which was too high, a failure that became evident when fisheries and survey data for 1997 became available. Korsbrekke *et al.* (2001), who compared the results of the traditional ICES stock assessments of cod with the results of assessments based on surveys only, concluded that the surveys had provided reliable and timely estimates of abundance for the stock in the period 1993–1999.

The reasons why the annual ICES-assessments tended to underestimate mortality rates and overestimate stock numbers in nearly all years between 1977 and 1998 are not fully known. A similar tendency was also found for other cod stocks (Korsbrekke *et al.* 2001). For Northeast Arctic cod, underreporting of catches may have contributed to the errors. In the early 1990s and early 2000s, the Norwegian Coastguard estimated that large quantities of cod were caught in addition to the quantities reported. Whatever the cause of the assessment error, the overestimation of stock size for a given year has the unfortunate effect that the stock size will be adjusted downwards in subsequent assessments, thus rendering agreed management strategies ineffective. The combined effects of too "optimistic" advice and even more optimistic agreed TACs, as well as catches exceeding those TACs, were that the yield from the stock during the first 25 years of Norwegian-Russian management was well below the stock's potential. However, the development of this stock has been far better than for most other stocks of cod in the world.

The most important result of the research and experiences described in this chapter is the implementation of the harvest control rule that was agreed by Norway and Russia in 2002 and approved by ICES in 2005. The rule, which includes precautionary considerations for the setting of annual TACs, will stabilize and enhance yields provided the regulations of the fisheries are satisfactorily monitored and enforced.

REFERENCES

Aglen, A., Nakken, O. 1997. Improving time series of abundance indices by applying new knowledge. Fisheries Research 30 (1997) 17–26.

Aglen, A., Drevetnyak, K., Sokolov, K. 2004. Cod in the Barents Sea (Northeast Arctic cod), a review of the biology and history of the fishery and its management. In: A. Bjordal, H. Gjøsæter, S. Mehl (Editors). Management strategies for commercial marine species in northern ecosystems. Proceedings of the 10th Norwegian-Russian Symposium, Bergen, August 2003. Institute of Marine Research, Bergen, and Polar Institute of Marine Fisheries and Oceanography, Murmansk, 2004.

Ajiad, A., Jakobsen, T., Nakken, O. 1999. Sexual difference in maturation of Northeast Arctic cod. Journal of Northwest Atlantic Fisheries Science, Vol. 25: 1–15.

Baranenkova, A.S. 1961. Soviet investigations on young cod in the Barents Sea. Annales Biologiques, Vol. 18: 105–107.

Berger, T.S. 1965. Peculiarities in the distribution of Cod in the Barents Sea and the Bear Island–Spitzbergen area. Annales Biologiques, Vol. 22: 82–86.

Bergstad, O.A., Jørgensen, T., Dragesund, O. 1987. Life history and ecology of the gadoid resources of the Barents Sea. Fisheries Research, 5: 119–161.

Beverton, R.J.H., Hylen, A., Østvedt, O.J. 1994. Growth, maturation and longevity of maturation cohorts of Northeast Arctic cod. ICES Marine Science Symposia, 198: 482–501.

Bogstad, B., Tjelmeland, S. (Eds.). 1992a. Interractions between fish populations in the Barents Sea. Proceedings of the Fifth PINRO–IMR Symposium, Murmansk, 12–16 August 1991. Institute of Marine Research, Bergen.

Bogstad, B., Tjelmeland, S. 1992b. A method for estimation of predation mortalities on capelin using a cod–capelin model for the Barents Sea. Proceedings of the Fifth PINRO–IMR Symposium, Murmansk, August 1991. Institute of Marine Research, Bergen.

Daan, N. 1994. Trends in North Atlantic cod stocks: a critical summary. ICES Marine Science Symposia, 198: 269–270.

Dahle, G. 1991. *Gadus morhua* L., populations identified by mitochondrial DNA. Journal of Fish Biology (1991) 38: 295–303.

Dannevig, G. 1951. Recaptures in the Barents Sea of cod tagged in Norwegian waters. Annuales Biologiques, Vol. 8: 16–17.

Dickson, R.R., Pope, J.G., Holden, M.J. 1974. Environmental influences on the survival of North Sea cod. In: The Early Life History of Fish, pp 69–80. Ed. by J.H.S. Blaxter. Springer-Verlag, Berlin. 765 pp.

Dingsør, G.E. 2001. Estimation of discards in the commercial trawlfishery for Northeast Arctic cod (*Gadus morhua* L.) and some effects on assessment. Cand.scient. thesis, University of Bergen, 2001.

Eggvin, J. 1937. Trekk av Nord-Norges oseanografi sett i sammenheng med torskefisket (Oceanographical conditions in North Norway connected with the cod fisheries). Report on Norwegian Fishery and Marine Investigations Vol. 5. no. 7: 33–46.

Ellertsen, B., Fossum, P., Solemdal, P., Sundby S. 1989. Relation between temperature and survival of eggs and first-feeding larvae of Northeast Arctic cod (*Gadus morhua* L.). Rapports et Procès-verbeaux des Réunions du Conseil International pour l'Exploration de la Mer, 191: 209–219.

Engesæter, S. 2002. The importance of ICES in the establishment of NEAFC. ICES Marine Science Symposia, 215: 572–581.

Engås, A. 1994. The effects of Trawl Performance and Fish Behaviour on the Catching Efficiency of Demersal Sampling Trawls. In: A. Fernø and S. Olsen (Editors). Marine Fish Behaviour in Capture and Abundance Estimation. Fishing News Books, Oxford.

Garrod, D.J. 1967. Population dynamics of the Arcto-Norwegian cod. Journal of Fisheries Research Board of Canada, 24(1): 145–190.

Garrod, D.J., Jones, B.W. 1974. Stock and recruitment relationships in the Northeast Arctic cod stock and the implications for management of the stock. Journal du Conseil International pour l'Exploration de la Mer, 36: 35–41.

Godø, O.R. 1984. Migration, mingling and homing of Northeast Arctic cod from two separated spawning grounds. In the proceedings of the Soviet-Norwegian symposium on Reproduction and Recruitment of Arctic Cod, Leningrad 26–30 September 1983, edited by O.R. Godø, S. Tilseth. Institute of Marine Research, Bergen, Norway, 1984.

Godø, O.R., Moksness, E. 1987. Growth and maturation of Norwegian coastal cod and Northeast Arctic cod under different conditions. Fisheries Research, 5: 235–242.

Godø, O.R. 2003. Fluctuations in stock properties of Northeast Arctic cod related to long-term environmental changes. Fish and Fisheries, 2003, 4. 121–137.

Hamre, J. 1988. Some aspects of the interrelation between the herring in the Norwegian Sea and the stocks of capelin and cod in the Barents Sea. ICES CM 1988/H: 42.

Hamre, J. 1991. Interrelation between environmental changes and fluctuating fish populations in the Barents Sea. In: Proceedings from an international symposium on long-term variability of pelagic fish populations and their environment, Sendai, Japan, 1989, pp. 259–270. Ed. by T. Kawasaki.

Heino, M., Dieckmann, U., Engelhard, G., Godø, O.R. 2004. Evolusjonære effekter av fisk (Evolutionary effects of fishing). In: K. Michalsen (Editor). Havets ressurser 2004, Fisken og havet, Særnr. 1–2004.

Hjort, J. 1903. Undersøgelser og iagttagelser. Årsberetning vedkommende Norges Fiskerier 1903: 33–35. (In Norwegian).

Hjort, J. 1908. Some results of the International Ocean Research. The Scottish Oceanographical Laboratory, Edinburgh, 40 pp.

Hjort, J. 1914. Fluctuations in the great fisheries of northern Europe reviewed in the light of biological research. Rapports et Procès-Verbaux des Réunions du Conseil International pour l'Exploration de la Mer, 20: 1–228.

Hylen, A. 1966. On the estimation of cod and haddock discarded by trawlers using different chafers. Appendix to Liaison Committee at ICES, Report to the North-East Atlantic Fisheries Commission, 1966.

Hylen, A. 1967. Discarding of fish in North-East Atlantic. ICES CM 1967/F:35.

Hylen, A., Smedstad, O. 1972. Norwegian investigations on young cod and haddock in the Barents Sea and adjacent waters 1970–1972. ICES CM 1972/F:38, 18 pp.

Hylen, A., Smedstad, O. 1974. Observations from the Barents Sea in spring 1973 on the discarding of cod and haddock caught in bottom and midwater trawls fitted with double cod ends. ICES CM 1974/F:45.

Hylen, A., Nakken, O., Sunnanå, K. 1986. The use of acoustic and bottom trawl surveys in the assessment of Northeast Arctic cod and haddock stocks. In: M. Alton (Editor). A Workshop on Comparative Biology, Assessment and Management of Gadoids from the North Pacific and Atlantic Oceans. Northwest and Alaska Fisheries Centre, Seattle, WA, pp. 473–498.

Hylen, A. 1987. Størrelsesfordeling til trålfanget torsk 1987. Notat til Fiskeridirektøren (Size distribution of cod caught in trawls in 1987. Note to the Director of Fisheries). (In Norwegian). Institute of Marine Research.

Hylen, A., Jacobsen, J.A. 1987. Estimation of cod taken as by-catch in the Norwegian fishery for shrimp north of 69°N. ICES CM 1987/G:34.

Hylen, A. 2002. Fluctuations in abundance of Northeast Arctic cod during the 20th century. ICES Marine Science Symposia, 215: 543–550.

ICES. 1971. Report of the Liaison committee of ICES to the North-East Atlantic Fisheries Commission 1971. Cooperative research report, Series B, 1971.

ICES. 1980. Reports of the ICES Advisory Committee on Fishery Management, 1980. Cooperative Research Report, 102.

Iversen, T. 1933. Some observations on fry in trawl catches in the Barents Sea. Rapport et Procès-Verbaux des Réunions, Vol. 85 (appendix 3 to North Eastern Area committee report): 3–6.

Iversen, T. 1934. Some observations on cod in northern waters. Fiskeridirektoratets Skrifter, Serie Havundersøkelser, 6 (8): 1–35.

Jakobsen, T. 1987. Coastal cod in northern Norway. Fisheries Research, 5: 223–234.

Jakobsen, T. 1992. Biological reference points for Northeast Arctic cod and haddock. ICES Journal of Marine Science, 49: 155–166.

Jakobsen, T. 1993. Management of Northeast Artic cod: past, present and future. In: Proceedings of the International Symposium on Management Strategies for Exploited Fish Populations, pp. 321–338. Eds: G. Kruse, D.M. Eggers, R.J. Marasev, C. Pautzke, T.J. Quinn. Alaska Sea Grant College Program, University of Alaska, Fairbanks, Alaska, USA. 825 pp.

Jakobsen, T. 1996. The relationship between spawning stock and recruitment for Atlantic cod stocks. ICES CM 1996.

Jakobsen, T., Korsbrekke, K., Mehl, S., Nakken, O. 1997. Norwegian combined acoustic and bottom trawl surveys for demersal fish in the Barents Sea during winter. ICES CM 1997/Y:17, 26 pp.

Jørgensen, T. 1990. Long-term changes in age at sexual maturity of Northeast Arctic cod (*Gadus morhua* L.). Journal du Conseil

International pour l'Exploration de la Mer, 46: 235–248.

Jørgensen, T. 1992. Long-term changes in growth of Northeast Arctic cod (Gadus morhua L. and some environmental influences. ICES Journal of Marine Science, 49: 263–277.

Korsbrekke, K., Mehl, S., Nakken, O., Pennington, M. 2001. A survey-based assessment of the Northeast Arctic cod stock. ICES Journal of Marine Science, 58: 763–769.

Mankevich, E.M. 1969. Structure of the stock of Arcto-Norwegian cod in 1969 according to age samples obtained off the north western coast of Norway. Annales Biologiques, Vol. 26: 123–125.

Marshall, T.C., Kjesbu, O.S., Yaragina, N.A., Solemdal, P., Ulltang, Ø. 1998. Is spawner biomass a sensitive measure of the reproductive and recruitment potential of Northeast Arctic cod? Canadian Journal of Fisheries and Aquatic Science, 55: 1766–1783.

Maslov, N. 1956. Soviet Investigations into the Biology of Gadoid Fish in the Barents Sea. Annales Biologiques, Vol. 13: 141–145.

Mehl, S. 1989. The Northeast Arctic cod stock's consumption of commercially exploited prey species in 1984–1986. Rapports et Procès-Verbaux des Réunions du Conseil International pour l'Exploration de la Mer, 88: 185–205.

Mehl, S. 1991. The Northeast Arctic cod stock's place in the Barents Sea ecosystem in the 1980s: an overview, Polar Research, 10(2): 525–534.

Mork, J., Ryman, N., Ståhl, G., Utter, F., Sundnes, G. 1985. Genetic variation in Atlantic cod (Gadus morhua L.): little divergence throughout the species range. Canadian Journal of Fisheries and Aquatic Sciences, 42: 1580–1587.

Møller, D. 1968. Genetic diversity in spawning cod along the Norwegian coast. Hereditas, 60: 1–32.

Møller, D. 1969. The relation between Arctic and coastal cod in their immature stages illustrated by frequencies of genetic characters. Fiskeridirektoratets Skrifter, Serie Havundersøkelser, 15: 220–233.

Nakken, O., Raknes, A. 1987. The distribution and growth of Northeast Arctic cod in relation to bottom temperatures in the Barents Sea, 1978–1984. Fisheries Research, 5: 243–252.

Nakken, O. 1994. Causes of trends and fluctuation in the Arcto-Norwegian cod stock. ICES Marine Sciences Symposia, 198: 212–228.

Nakken, O., Sandberg, P., Steinshamn, S.J. 1996. Reference points for optimal fish stock management. A lesson to be learned from the Northeast Arctic cod stock. Marine Policy, Vol. 20, No. 6, pp. 447–462.

Nakken, O. 1998. Past, present and future exploitation and management of marine resources in the Barents Sea and adjacent areas. Fisheries Research, 37 (1998): 23–35.

Nakken, O. 2002. Understanding environmental controls on fish stocks and progress towards their inclusion in fish stock assessment. ICES Marine Sciences Symposia, 215: 247–255.

Olsen, S. 1968. Some results of the Norwegian capelin investigations 1960–1965. Rapports et Procès-Verbaux des Réunions du Conseil International pour l'Exploration de la Mer, 158: 18–23.

Ottersen, G., Loeng, H., Raknes, A. 1994. Influence of temperature variability on recruitment of cod in the Barents Sea. ICES Marine Sciences Symposia, 198: 471–481.

Ottersen, G., Sundby, S. 1995. Effects of temperature, wind and spawning stock biomass on recruitment of Arcto-Norwegian cod. Fisheries Oceanography 4:4, 278–292.

Ottersen, G. 1996. Environmental impact on variability in recruitment, larval growth and distribution of Arcto-Norwegian cod. PhD thesis. University of Bergen, Norway. 136 pp.

Ottersen, G., Michalsen, K., Nakken, O. 1998. Ambient temperature and distribution of Northeast Arctic cod. ICES Journal of Marine Science, 55: 67–85.

Ottestad, P. 1942. On periodical variations in yield of the great sea fisheries and the possibility of establishing yield prognosis. Fiskeridirektoratets Skrifter, Serie Havundersøkelser (Report of the Norwegian Fisheries and Marine Investigations), 7 (5): 3–11.

Petersen, C.G.J. 1903. What is overfishing? Journal of Marine Biological Association, 6: 587–594.

Planque, B., Fox, C.J. 1998. Interannual variability in temperature and the recruitment of Irish Sea cod. In: Report of the ICES/GLOBEC Workshop on Application of Environmental Data in Stock Assessment, pp. 55–61. ICES CM 1998/C:1. 97 pp.

Ponomarenko, V.P. 1973. On a probable relation between age composition of spawning stock and abundance of the year classes of cod in the Barents Sea. Rapports et Procès-Verbaux des Réunions du Conseil

International pour l'Exploration de la Mer, 164: 69–72.

Ponomarenko, V.P., Ponomarenko, I.Y. 1975. Consumption of the Barents Sea capelin by cod and haddock. ICES CM 1975/F: 10.

Rollefsen, G. 1930. Observations on cod eggs. Rapports et Procès-Verbaux des Réunions du Conseil International pour l'Exploration de la Mer, 65: 31–34.

Rollefsen, G. 1932. The susceptibility of cod eggs to external influences. Journal du Conseil, 7 (3): 367–373.

Rollefsen, G. 1933. The otoliths of the cod. Preliminary report. Fiskeridirektoratets Skrifter, Serie Havundersøkelser (Report of the Norwegian Fisheries and Marine Investigations), 4 (3): 3–14.

Rollefsen, G. 1935. The spawning zone in cod otoliths and prognosis of stock. Fiskeridirektoratets Skrifter, Serie Havundersøkelser (Report of the Norwegian Fisheries and Marine Investigations), 4: 3–10.

Rollefsen, G. 1937. Aldersundersøkelser (Age analysis of the 1937 catch of cod). Report in Norwegian Fishery and Marine Investigations, 5 (7): 23–31. (In Norwegian).

Rollefsen, G. 1938. Changes in mean age and growth-rate of the year-classes in the Arcto-Norwegian stock of cod. Rapports et Procès-Verbaux des Réunions du Conseil International pour l'Exploration de la Mer, 100: 37–44.

Rollefsen, G. 1946. Den utenlandske tråling og torskebestanden i de nordlige farvann. Fiskeridirektoratets Småskrifter, 1946 (11): 1–17.

Rollefsen, G., Tåning, Å.V. 1948. Enquiry into the problem of climatic and ecological changes in northern waters. Rapports et Procès-Verbaux des Réunions du Conseil International pour l'Exploration de la Mer, 125: 8.

Rollefsen, G. 1949. Fluctuations in two of most important stocks of fish in northern waters, the cod and the herring. Rapports et Procès-Verbaux des Réunions du Conseil International pour l'Exploration de la Mer, 125: 33–35.

Rollefsen, G. 1954. Observations on the cod and cod fisheries of Lofoten. Rapports et Procès-Verbaux des Réunions du Conseil International pour l'Exploration de la Mer, 136: 40–47.

Rollefsen, G. 1966. Norwegian fisheries research. Fiskeridirektoratets Skrifter, Serie Havundersøkelser (Report of the Norwegian Fisheries and Marine Investigations), 14 (1): 1–36.

Rothschild, B.J., Osborn, T.R. 1988. Small-scale turbulence and plankton contact rates. Journal of Plankton Research, 10: 465–474.

Sars, G.O. 1879. Inberetninger til Departementet for det Indre fra professor G.O. Sars om de af ham anstillede undersøgelser angaaende saltvandsfiskeriene. Berg og Ellefsens Bogtrykkeri, Christiana: 221 pp. (In Norwegian).

Schwach, V. 2000. Havet, fisken og vitenskapen. Fra fiskeriundersøkelser til Havforskningsinstituttet 1860–2000. Institute of Marine Research, Bergen: 405 pp. (In Norwegian).

Serebryakov, V.P. 1991. Predicting year-class strength under uncertainties related to survival in the early life history of some North Atlantic commercial fish. NATO Scientific Council Studies, 16: 49–55.

Solemdal, P. 1997. Maternal effects – a link between the past and the future. Journal of Sea Research, 37: 213–227.

Solemdal, P., Serebryakov, V. 2002. Cooperation in marine research between Russia and Norway at the dawn of the 20th century. ICES Journal of Marine Sciences, 215: 73–86.

Stransky, C., Baumann, H., Fevolden, S.E., Harbitz, A., Høie, H., Nedreaas, K.H., Salberg, A.B., Skarstein, T.H. 2007. Separation of Norwegian coastal cod and Northeast Arctic cod by otolith morphometry. ICES Council Meeting 2007/L:10.

Sund, O. 1927. The Arcto-Norwegian cod stock. Journal du Conseil International pour l'Exploration de la Mer, 2 (2): 161–169.

Sund, O. 1930. II. The renewal of a fish population studied by means of measurement of commercial catches. Example: the Arcto-Norwegian cod stock. In: Fluctuations in the Abundance of the Various Year-Classes of Food Fishes. Pp. 10–17. Rapports et Procès-Verbaux des Réunions du Conseil International pour l'Exploration de la Mer, 65, 115 pp.

Sund, O. 1933. Torskebestanden (The stock of cod). Årsberetning vedk. Norske fiskerier 1933 (Annual report Norwegian fisheries), 84–90.

Sund, O. 1936. The fluctuations in the European stocks of cod. Rapports et Procès-Verbaux des Réunions du Conseil International pour l'Exploration de la Mer, 101 (3): 1–18.

Sundby, S., Fossum, P. 1990. Feeding conditions of Arcto-Norwegian cod larvae compared with the Rothschild-Osborn theory on small-scale turbulence and plankton contact rates. Journal of Plankton Research, 12: 1153–1162.

Sundby, S. 2000. Recruitment of Atlantic cod stocks in relation to temperature and advection of copepod populations. Sarsia, 85: 277–298.

Sundby, S., Nakken, O. 2005. Spatial shifts in spawning habitats of Arcto-Norwegian cod induced by climate change. GLOBEC International Newsletter, 11(1): 26.

Sætersdal, G. 1962. Torsk, sei og hyse (cod, saithe and haddock). In: G. Rollefsen (Editor), Havet og våre fisker (The sea and our fishes). J.W. Eides Forlag, 1962.

Sætersdal, G., Hylen, A. 1964. The decline of the skrei fisheries. Fiskeridirektoratets Skrifter, Serie Havundersøkelser, 13(7): 56–69.

Sætersdal, G. 1989. Fish resources research and fishery management: A review of nearly a century of experience in the Northeast Atlantic and some recent global perspectives. Journal du Conseil International pour l'Exploration de la Mer. 46 5–15, 1989.

Sætersdal, G. 1993. Fisheries research and fisheries management, historical perspectives and some current challenges. Keynote address at the 20th session of Committee of Fisheries, FAO. Rome, 15–19 March 1993.

Sætersdal, G., Loeng, H. 1987. Ecological adaptation of reproduction in Northeast Arctic cod. Fisheries Research, 5: 253–270.

Sætre, R. 2004. Scientific research in the Norwegian Sea: Background and history. In: The Norwegian Sea Ecosystem, Ed: H.R. Skjoldal, Tapir Academic Press, Trondheim 2004.

Tjelmeland, S., Bogstad, B. 1998. MULTSPEC – a review of a multispecies modelling project for the Barents Sea. Fisheries Research, 37: 127–142.

Trout, G.C. 1955. English Cod Tagging in 1954 and 1955. Annales Biologiques, 12: 138–140.

Trout, G.C. 1957. The Bear Island cod: migrations and movements. Fishery Investigation, London, Series 2, 21(6), 51 pp.

Tåning, Å.V. 1949. Fluctuations in Fish Populations owning to Climatic Changes. In: UNSCCUR proceedings: Wildlife and Fish Resources, 1949.

Ulltang, Ø. 1987. Potential gains from improved management of the Northeast Arctic cod stock. Fisheries Research, 5: 243–252.

Ulltang, Ø. 1996. Stock assessment and biological knowledge: can prediction uncertainty be reduced? ICES Journal of Marine Sciences, 53: 659–675.

Wiborg, K.F. 1957. Factors influencing the size of the year-classes in the Arcto-Norwegian tribe of cod. Fiskeridirektoratets Skrifter, Serie Havundersøkelser, 11(8): 1–24.

Woodhead, A.D. 1959. Variations in the activity of the thyroid gland of the cod, *Gadus callaris* L., in relation to its migrations in the Barents Sea. II. The "dummy run" of the immature fish. Journal of Marine Biological Association, UK, 38: 417–422.

Øiestad, V. 1994. Historical changes in cod stocks and cod fisheries: Northeast Arctic cod ICES Marine Sciences Symposia, 198: 17–30.

CHAPTER 6

The Barents Sea 0-group surveys; a new concept of pre-recruitment studies

Olav Dragesund, Arvid Hylen, Steinar Olsen and Odd Nakken

6.1 Introduction

The most frequently used method of estimating relative year-class strength in marine fish populations is to compare the frequency distributions of the different age groups in the exploited stocks. For fisheries management, however, the ability also to assess and thereby predict recruitment fluctuations at an earlier stage has always been important. The first question in this connection is then: at what age or stage in the life history is the strength of a year-class determined?

While survival at the egg and larval stages has long been found to vary widely due to a multitude of other causes than fluctuations in egg numbers spawned, in the case of post-larval fish fry evidence has been presented of a proportional relationship between the abundance of 0-group fish of a particular year-class and the subsequent abundance of the same year-class at greater ages. In 1963, Olav Dragesund and Steinar Olsen initiated a study of the distribution and abundance of 0-group commercial fish species in the Barents Sea region. They assumed as a working hypothesis for their investigation that such a proportional relationship exists, and suggested that year-class strength might be estimated by measuring the echo-abundance of 0-group fish (Dragesund and Olsen 1965). From their own previous investigations (Dragesund and Hognestad 1960, 1962; Olsen 1960) they knew that from August to October, fish fry of various species, originating from the spawning fields farther south, are abundant in the surface layers off the coast of northern Norway and in the Barents Sea (Figure 6.1), and with suitable echo-sounders can be detected as distinct scattering layers. At the end of the autumn, the fish fry had been found to be concentrated along the frontiers between the cold and warm water masses in the area from Spitsbergen to Bear Island and further to the east and south in the central and south-eastern parts of the Barents Sea. Concentrations were also observed along the coast of northern Norway, especially at the entrances to fjords. In late autumn, 0-group cod, haddock, and to some extent also herring and capelin, descend and settle near the bottom, hence during the winter months fish fry are found more rarely in the surface layers.

During the late summer and autumn, fish fry of several important commercial species thus occur pelagically off the north Norwegian coast and in the Barents Sea, and it was suggested that their distribution and abundance might be charted

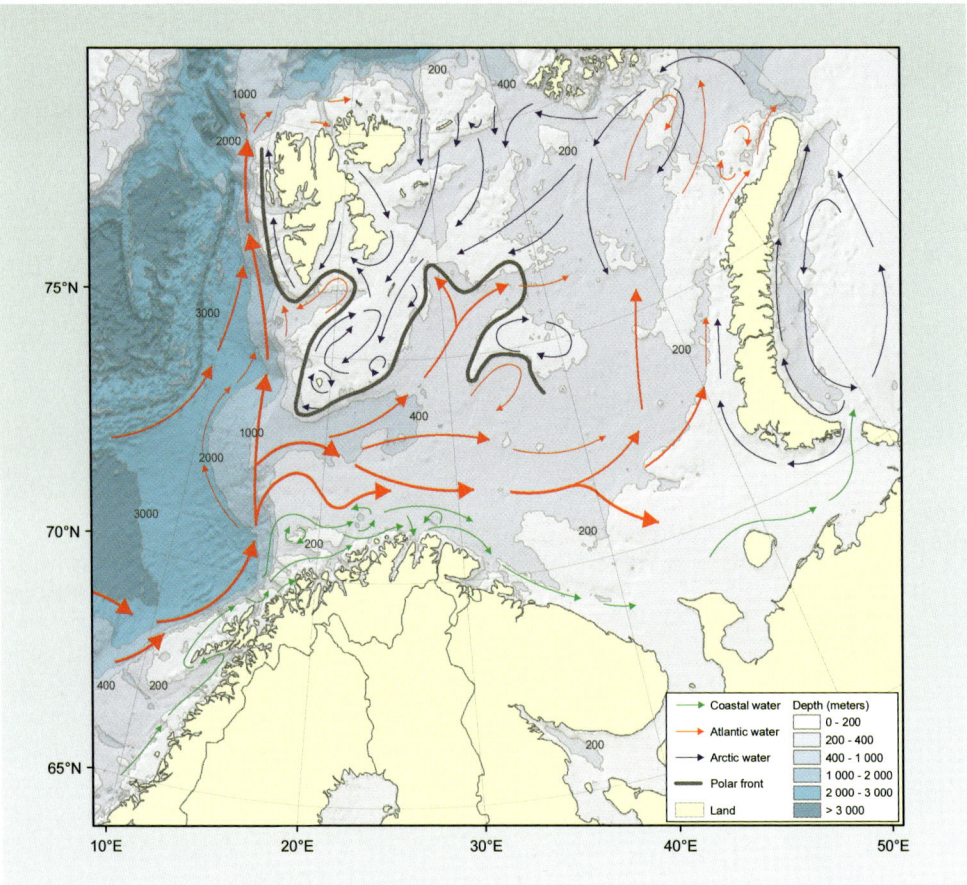

Figure 6.1
Ocean currents in the area, showing the transport routes of eggs and larvae from the spawning grounds off the Norwegian coast into the Barents Sea.

and measured by a combination of echo surveying and experimental fishing using pelagic trawl and purse seine net. The success of such investigations would depend on fulfilment of certain requirements:
1. reasonably complete and accurate charting of the vertical and horizontal distribution of the sound scatterers
2. easy and reliable identification of the sound scatterers
3. exact measurements of the echo signals received (echo abundance)
4. knowledge of the relationship between the amount of scatterers and echo strength, and how this relationship is influenced by depth, species, size, density and fish behaviour.

Most of these requirements were regarded as mainly requiring sufficient research effort and vessel time to fulfil. However, since previous attempts at quantifying echo recordings of scattering layers, and other forms of fish distribution had been based on subjective evaluations, the main challenge was to develop instrumentation and methods that could provide accurate and unbiased numerical estimates of the amount of echo signals received. In August 1963, instrument chief Ingvar Hoff, with some initial advice from Simrad, succeeded in engineering an electronic analogue echo-integrator that facilitated the summing of all signal voltages generated by the echo-sounder within a selected time interval (i.e. depth range). This success and the first echo-integrator chart of fish fry (mainly cod) were presented as a note to the ICES Fish Abundance Measurement Symposium in Madrid 1963 (Gulland 1964), but a full description of the echo-integrator was not published until two years later (Dragesund and Olsen 1965). Figures 6.2 and 6.3 show examples from the first use of the instrument.

In 1964, further developments and tests of sampling gear and methods for surveying 0-group fish were carried out, and at the 1964 statutory meeting of ICES, the Herring Committee made the following recommendation: "The Committee recommends strongly that Norwegian and Soviet research vessels should undertake joint surveys of the distribution of the early stages of herring in the

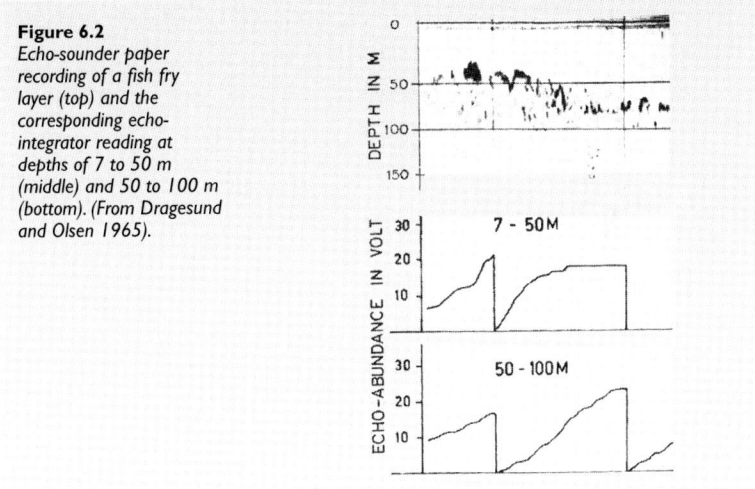

Figure 6.2
Echo-sounder paper recording of a fish fry layer (top) and the corresponding echo-integrator reading at depths of 7 to 50 m (middle) and 50 to 100 m (bottom). (From Dragesund and Olsen 1965).

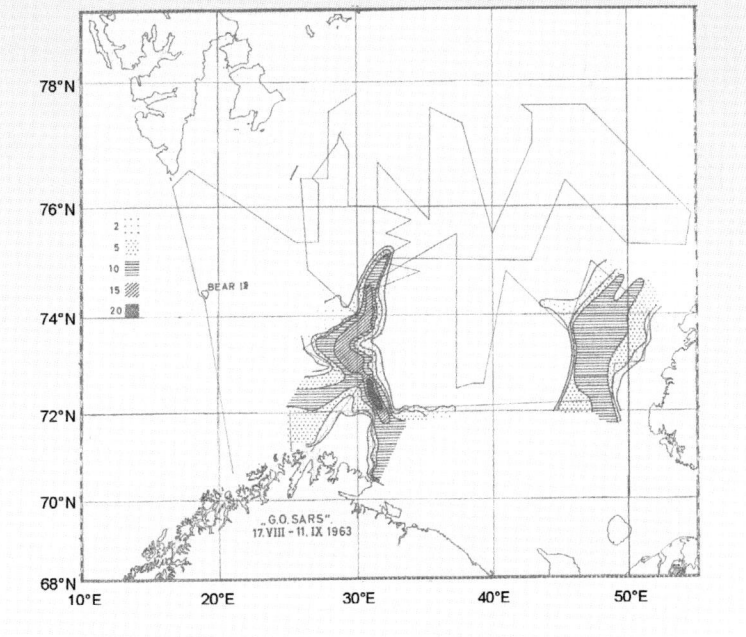

Figure 6.3
The first echo-integrator charting. Echo-abundance distribution as determined during the survey from 17 August to 11 September 1963. Equal levels of abundance are indicated by isolines. (From Dragesund and Olsen 1965).

Yuriy K. Benko, Amirova Elizaveta Arkad'evna, Alexander S. Seliverstov (all from PINRO) and Steinar Olsen (IMR) at an 0-group survey planning meeting.

Eastern Norwegian Sea and the Barents Sea". The general programme for this survey was discussed by Soviet and Norwegian scientists in Moscow in May 1965, and it was agreed that the aims should be to investigate the distribution and abundance, not only of herring, but also of other commercial species of fish and to make hydrographic observations (ICES 1965). Accordingly, since 1965 the Institute of Marine Research and the Polar Research Institute of Marine Fisheries and Oceanography (PINRO), Murmansk, have carried out every year in August–September, without interruption, extensive, multi-ship 0-group surveys of the Barents Sea region, in 1966–1973 also with the participation of the Marine Laboratory, Lowestoft, UK.

6.2 The initial period of the Barents Sea 0-group surveys

In addition to the preliminary reports of the investigations presented at the ICES meetings each year (ICES 1965, 1966, 1967 and 1969), an account of the first four years' surveys and their most important results was published as an ICES Cooperative Research Report (Dragesund 1970). Here we shall mainly recount the organisation and survey methodology employed during these years and the general experience gained.

Detailed programme planning was carried and agreed on every year immediately prior to the start of the surveys during calls at Murmansk by the research vessels. The survey grid aimed to cover almost the entire area of distribution of fish fry of commercial species north of the Lofoten Islands. Individual research vessel tracks were arranged so that they were to operate in pairs consisting of one

Norwegian and one Soviet vessel, steaming more or less parallel courses 30–40 miles apart. In the years when the UK research vessel participated, it became possible to survey the whole area in more detail and to extend it further west into the Norwegian Sea. A typical grid pattern is shown in Figure 6.4.

At selected grid lines, hydrographic observations were made in order to relate the distribution of organisms in the scattering layer to the principal hydrographic features. During the surveys, radio communication between the participating vessels was maintained twice a day to report the observations made. At the end of the surveys the vessels met, all data and observations were combined, and preliminary reports were prepared for submission to ICES.

During the survey, continuous records of the pelagic scattering layer were made. In order to ensure comparability of the results, the participating vessels were equipped with the same echo-sounder, and intership acoustic calibrations were carried out by comparing results obtained from the same area. These showed that the methods employed gave fairly consistent agreement between vessels as to the location and density of traces.

The echo abundance was gauged from the paper recordings on a subjective scale from 0 to 4. The corresponding measurements with the one and only echo-integrator available at the time showed that these subjective estimates

Figure 6.4
Survey routes and grid of stations during the 0-group survey 1967.

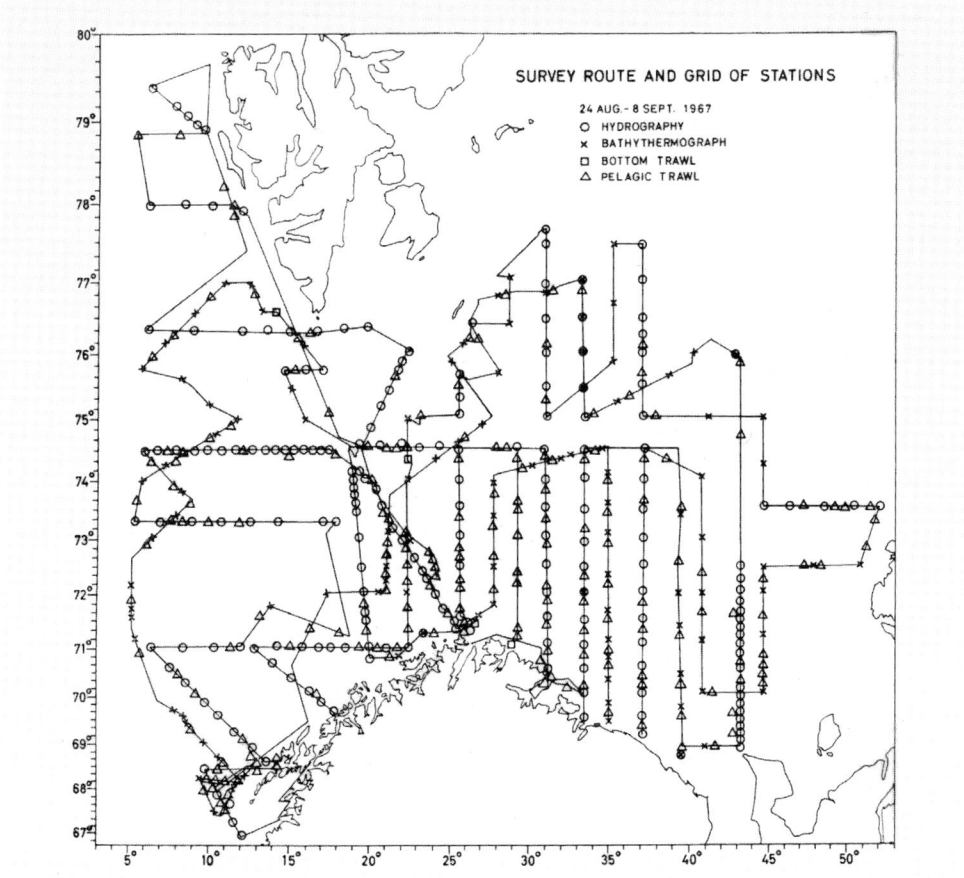

were too coarse and rather inaccurate at high densities. Further experimental work on the influence of diurnal variations in the distribution of the scatterers was also found to be necessary before more precise quantitative measurements could be made. For identification and sampling, pelagic trawl hauls were made at appropriate intervals in areas and at depths where sound scatterers were being recorded, but usually not more than 40 miles apart. In addition, some control bottom trawl hauls were made to check whether some 0-group fish had already migrated out of the pelagic scattering layer.

Identification of the sound scatterers was also partly facilitated by the characteristics of the echo recordings made. 0-group cod and haddock are usually found deeper than herring, and during the day they show a different schooling pattern. Cod and haddock do not form well-defined schools, but appear as layers of more or less discrete concentrations. Herring occur in small concentrated schools, which are easily detected by sonar. This feature distinguishes 0-group herring from redfish, which are often observed in the same depth range as herring, but do not usually form well-defined schools. 0-group capelin and long rough dab have lower target strengths than fry of redfish, herring, cod and haddock, and recordings of these can usually be easily separated from those of other species. Especially in situations where several species occurred together it was inescapable, however, to supplement acoustic identification with fishing experiments. The catching gear used by all participating vessels was a fine-meshed pelagic trawl, and the fish-capture power of the trawls was tested by intership comparisons carried out before the surveys. The depth of trawling was checked by a depth recorder attached to the trawl.

The abundance of the different species recorded in the scattering layer showed variations from one year to the next. Redfish and capelin were rather abundant in all years, whereas herring, cod and haddock occurred in very low abundance every year. The general impression gained from the surveys was that the total abundance of 0-group fish in the scattering layer did not vary to the same extent as the abundance within species from year to year. However, this had to be verified by estimating more thoroughly the echo-abundance of the different 0-group fish in the scattering layer. Thus, the importance of a more accurate measurement of echo-abundance was stressed.

The results obtained during the first four years of surveys were judged by the participating scientists to be of sufficient interest to warrant continuation of the work. A joint survey ought therefore to be repeated on an annual basis, at least until separate assessments of the strength of some of the year-classes studied as 0-group fish could be made from the catches of the commercial fisheries. Since the main shortcoming of the technique used appeared to be related to identifying the echo-recordings, it was also recommended that high priority should be given to efforts to improve identification and sampling methods.

6.3 The subsequent 30 years
Vessels, gears and trawling procedure
More than 30 vessels were used in the survey from the start in 1965 until 2000, and three to six vessels participated every year. Those used in the first five years were all built as side-trawlers, some of them having arrangements so that the pelagic sampling trawl could be operated from the stern. Between 1970 and 1985, all side-trawlers were replaced by bigger and better-equipped stern-trawlers,

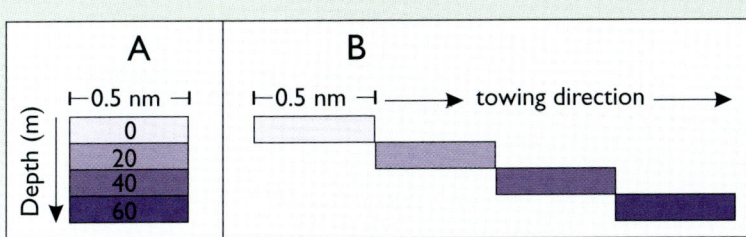

Figure 6.5
Trawling procedure adopted in 1981. A) Schematic representation of a trawl haul including four depth steps. B) How the haul is performed. (From Stensholt and Nakken 2001).

capable of operating larger trawls and with instrumentation for monitoring and controlling trawling performance. Before 1985, the trawls used varied considerably in size between vessels according to their size and propulsion power. From 1985 and onwards, all vessels have used identical trawls with a rectangular mouth opening of about 20 x 15 metres (see Nakken and Raknes 1996). This "standard" trawl is a commercial type of trawl designed to catch capelin. In the early 1990s, its performance and capture efficiency was thoroughly studied and compared with that of a trawl specially designed to catch 0-group fish. (Godø et al. 1993; Godø and Valdemarsen 1993; Valdemarsen and Misund 1995; Hylen et al. 1995). These studies showed that the catching efficiency of both trawls increased with increasing density of 0-group fish (cod and haddock) and that the length distributions obtained with the standard trawl were highly skewed compared with those provided by the special trawl; the standard trawl catching much fewer small-sized 0-group cod and haddock than the "special" trawl.

These differences in capture efficiency over the length and density ranges of the 0-group, will bias the comparability of the abundance indices over years. In years with low abundance and small-sized 0-group cod and haddock they will be underestimated as compared with years with higher abundance and larger size of 0-group individuals. Until 1980, pelagic trawl hauls were made at about every 30–40 nautical mile distance sailed, as well as when the characteristics of the echo-recordings of the 0-group scattering layer changed. Trawling depths were decided from the appearance of specific layers on the echo-sounder in order to sample the layer satisfactorily. When no recordings were available, the trawl was towed with the headline at, or as close as possible to the surface. Towing distance was 1 nautical mile at a speed of 3 knots. From 1981 and onwards, a new trawling procedure was adopted in order to ensure that the depth range of the 0-group was covered (Figure 6.5). Now the trawl was towed in a stepwise manner in each haul with the headline kept at each of the following depths: 0 m, 20 m, 40 m, and also at 60 and even 80 m when the echo recordings indicated the necessity of including these depths. The towing time at each depth was 10 minutes, equivalent to a distance of 0.5 nautical mile at a speed of 3 knots.

Establishing indices of abundance
The main aim of the survey was to obtain a measure of the annual abundance of each species of 0-group fish, a measure that could be used both to forecast the recruitment of fish to the fishery some years later and for further studies of

recruitment mechanisms in the stocks. In the first five or six years, the year-class strengths of the different species were stated to be poor, average or strong on the basis of a joint inspection of the distribution maps of echo-recordings and catch numbers. No calculations of numerical values representing the abundance of year-classes were attempted in these years. The subjective grading into poor, average or strong year-class strength depended heavily on the experience of the participating scientists. When it was decided in 1970 to continue the survey into future, the need for a numerical measure, i.e. an index of abundance, became pressing.

As mentioned above, the density of scatterers was estimated by visual grading of the echo-sounder paper recordings and classified on a subjective scale from 0 to 4 according to the following code:

Code number:	0	1	2	3	4
Classification:	No recording	Very scattered	Scattered	Dense	Very dense

During the cruises, the classifications for each nautical mile sailed were performed. The distributions of 0-group fish were plotted on maps in terms of three density grades; absent, scattered and dense recordings, this grading being partly based on visual inspection of the echo-recordings and partly on the quantity of fish caught in the trawl hauls. Examination of the distribution maps indicated that the criteria used to discriminate between scattered and dense had varied from year to year. The following method for calculating abundance indices was therefore adopted in 1972 (Haug and Nakken 1977): A certain catch rate (number of fish per distance trawled) was used to distinguish between scattered and dense concentrations in the distribution maps. This number was estimated by examining the catches containing only one species and comparing the catch rate with the corresponding visual gradings of the echo trace. In this way species-specific catch rates for distinguishing between scattered and dense were arrived at. Distribution maps were then made in three density grades as before, but now using the estimated species-specific catch rate to discriminate between scattered and dense concentrations. Finally, the total abundance index for each species was estimated directly from the map using the formula:

Total index = Area of low catch rate + 10 x Area of high catch rate

The ratio between fish densities classified as dense (codes 3 and 4) and scattered (codes 1 and 2), i.e. the number 10 in the formula, was based on echo-integrator readings (Haug and Nakken 1977) as well as on purse-seine catches of 0-group herring in the early 1960s (Dragesund 1970). This way of computing abundance indices often referred to as the "area index", became one of two standard methods used for the 0-group survey data.

The other index, the logarithmic one that was derived more directly from the trawl catch rates, was introduced in the early 1980s (Randa 1984). The entire area was now divided into strata, and mean values and variances of catch rates were computed for each stratum as well as for the entire area. The method, which allowed confidence limits to the indices to be estimated, came into regular use in 1982.

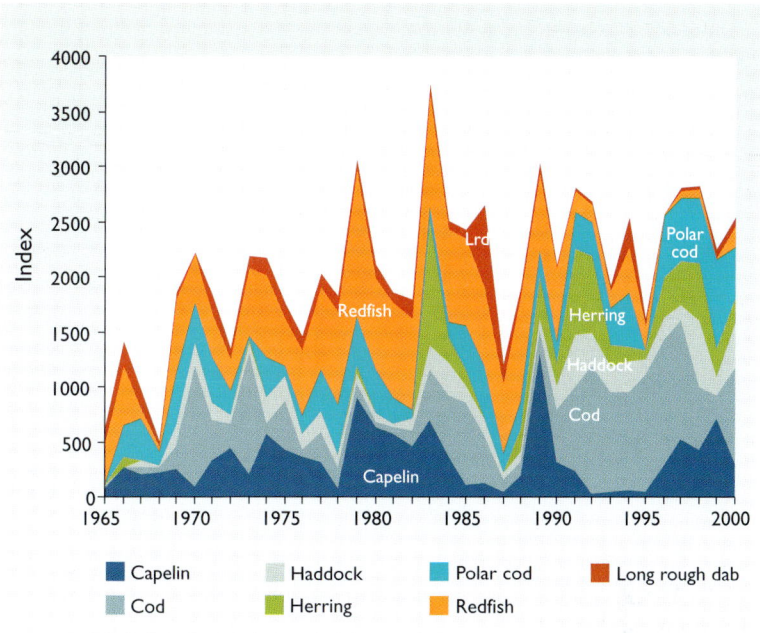

Figure 6.6
Annual abundance (area indices) of 0-group fish in the Barents Sea and adjacent waters 1965–2000. (Source ICES 2003)

Figure 6.6 shows the time series of the area indices for the main species (ICES 2003). 0-group herring were virtually absent in the survey data before 1983, and indices of herring abundance were computed later, when it was evident that the stock would recover (Toresen 1985).

As mentioned above, the size of the trawls used increased until 1984, a factor that produced an increase in the overall catching efficiency of the survey and thus a bias in the index values. In 1996, the area indices for cod, haddock and redfish prior to 1985 were corrected for this change in catching efficiency, in order to obtain a time series with comparable values over the entire period (Nakken and Raknes 1996).

Acoustic or/and catch based abundance indices?
As stated in the introduction to this chapter, the intention at the start of the survey was to establish indices of abundance based primarily on acoustic recordings, and to use the trawl catch rates more as a support. From the description above of the establishment of the indices, it is evident that developments in the 1970s went in the opposite direction. The area index established in the early 1970s (Haug and Nakken 1977) was partly based on acoustic recordings and partly on catch rates in trawl hauls, while the logarithmic index established in the early 1980s was exclusively based on catch rates (Randa 1982). However, during the 1970s, echo-integrators came into routine use in the Institute of Marine Research's survey activity, including the 0-group survey. Estimates of abundance based on acoustic surveys were worked out annually for a number of species (capelin, cod, haddock, blue whiting) so why was the abundance of 0-group species not estimated acoustically? In order to estimate the density of a given species from echo integration data, the species must either be the only contributor to the recording, or in the case of

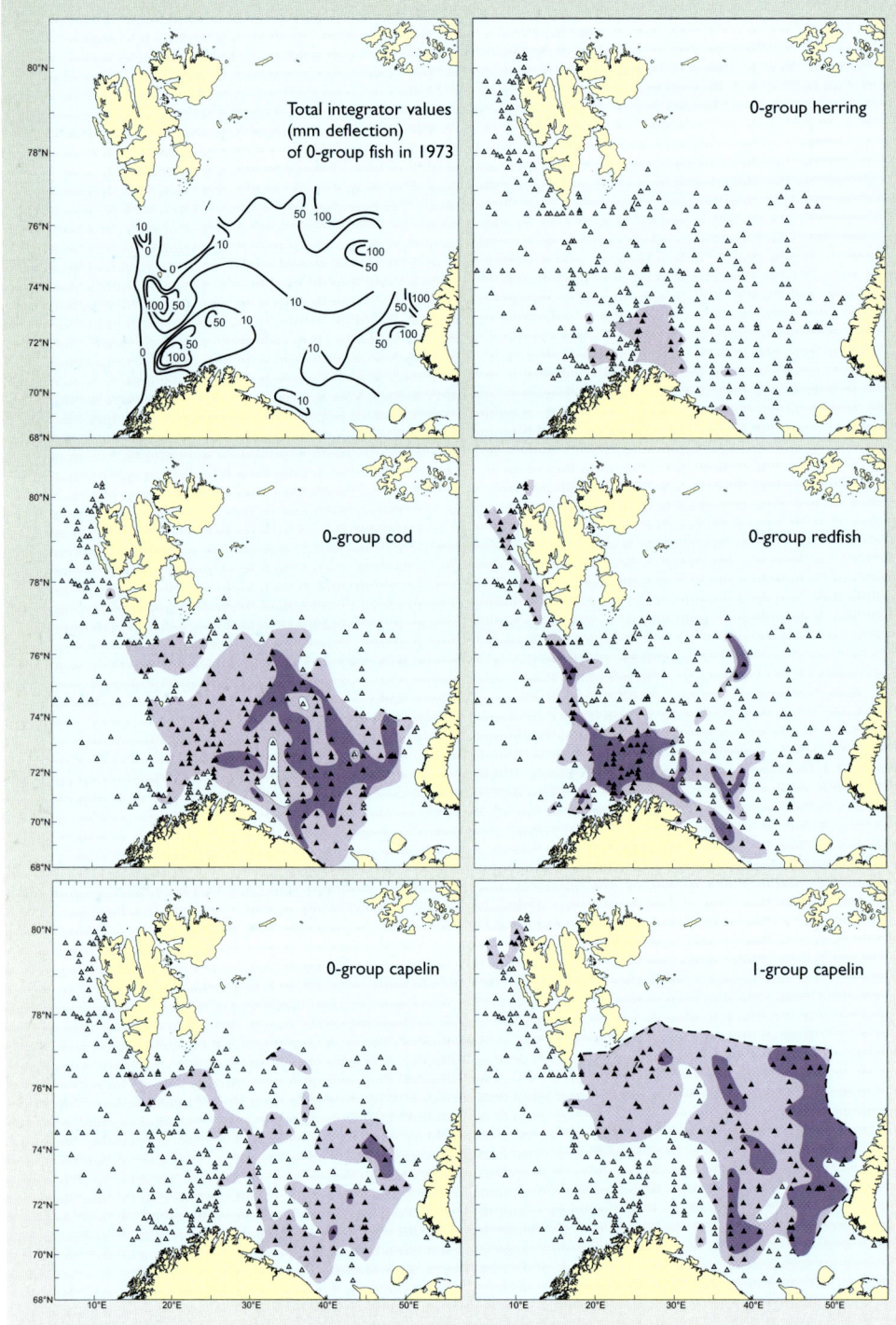

Figure 6.7 *Distribution of total integration values (all species) and catch rates for some species from the 1973 0-group survey.*

mixed recordings with contributions from several species, the species-specific contribution must be evaluated by partitioning the total integration value, for example, according to the composition of the trawl catches. Usually most trawl hauls from the 0-group scattering layer included many species of 0-group fish in addition to various species of jellyfish as well as amphipods and krill. In years when 0-group cod and/or 0-group herring were very abundant, it was possible to obtain acoustic estimates of the abundance (see later). However, in most years, and particularly throughout the 1970s when 0-group herring were absent, most year-classes of cod weak, and the 0-group scattering layer dominated by 0-group redfish and 0- and 1-group capelin, the partitioning of the total 0-group integration values became extremely uncertain. Figure 6.7 shows some distribution maps from the 1973 survey. The high echo abundance (integration values) in the western areas appeared to be associated with the catch distributions of 0-group redfish and herring, while those in the eastern areas were related to 0-group polar cod and 1-group capelin, which quite often appeared in the 0-group trawl hauls.

The hardware and software developments for post processing of acoustic data throughout the late 1980s enabled a more detailed and reliable assignment of echo integration values to species than previously. In the 1990s, when 0-group redfish became scarce and 0-group cod and herring were abundant in several successive years, acoustic estimates of abundance were computed for both of these species (Toresen and de Barros 1995; Nakken et al. 1995; Hylen 1997). The acoustic estimates of 0-group cod arrived at, were from five to 25 times as high as the numbers established by Sundby et al. (1989), which were based on the logarithmic abundance index assuming trawl catching efficiencies of 0.25–1.0. However, compared with estimates at later stages of the same year-classes (1-, 2- and 3-year-old fish) the numbers of 0-group arrived at from the acoustic recordings seemed reasonable, and it was concluded that these estimates of cod and haddock (Hylen 1997) and herring (Toresen et al. 1998) provided useful information on the absolute abundance of year-classes and hence on the natural mortality rates that these experienced at juvenile life stages (Tables 6.1 and 6.2).

6.4 Main general results

The results of the Barents Sea 0-group survey represent the longest unbroken time series of the spatial distribution of pelagic fish fry in this region, and the data have been widely used in studies of abundance, distribution (vertical and

Table 6.1 Northeast Arctic cod. Numbers at age (in billions) in the year-classes 1991–1994. "0" is the acoustic estimate of abundance from the 0-group survey. 1–3 are the estimates from the catch-at-age analyses when cannibalism has been accounted for. (From Hylen 1997).

Year-classes	Age (years)			
	"0"	1	2	3
1991	81	3,4	1,8	0,93
1992	107	2,8	1,8	0,72
1993	81	12	1,5	0,45
1994	99	32	2,4	

Table 6.2 Norwegian spring-spawning herring. Abundance (billion individuals) at age of small herring (Clupea harengus L.) in the period 1983–1994. (From Toresen et al. 1998).

Year-classes	Age (years)				Total
	"0"	1	2	3	
1983	33	0	0	0	33
1984	21	21	0	0	42
1985	47	1	20	0	68
1986	0	0,5	0,2	3	4
1987	0	0	0	0	0
1988	36	0	0	0	36
1989	19	2	0	0	21
1990	62	4	0	0	66
1991	101	24	5	0	130
1992	275	33	14	6	328
1993	100	103	26	0	229
1994	0	7	59	18	84

horizontal), behaviour and growth of the 0-group species, especially cod and herring, in relation to the environment (Beltestad *et al.* 1975; Loeng *et al.* 1995; Nakken 1994; Ottersen *et al.* 1994; Ottersen and Sundby 1995; Ottersen and Loeng 2000; Helle *et al.* 2000; Stensholt and Nakken 2001). Most of the 28 papers presented and discussed at the sixth IMR-PINRO symposium in June 1994 dealt with investigations that included data from the 0-group survey used to elucidate the dynamics of many fish stocks in the area (Hylen 1995).

Figures 6.8 and 6.9 show the "average" geographical distribution and the annual mean temperature of the habitats (ambient temperature) of the various 0-group species. As was to be expected, herring, redfish and the gadoids are the typical "warmwater" species, while polar cod are found in colder waters. It should be noted that the interannual variations in mean ambient temperatures (Figure 6.9) are, for most species, greater than the corresponding temperature variations in the Kola section, a finding that calls for caution when relating changes in biological variables to temperature changes at fixed stations or sections.

One of the main objectives of the survey was to provide measures of future recruitment to the fisheries. Has this objective been met? For year-classes of cod prior to the mid-1980s, a fair correlation existed between the 0-group index and stock numbers at age 3 from the catch-at-age analysis (Randa 1982, 1984), and the index was thus used directly to predict future recruitment (age 3) in the stock assessment for some years. But when the capelin stock, the main food source for cod in this region, collapsed in the mid 1980s, increased cannibalism in cod caused the established relationship between the 0-group index and numbers at age 3 to break down. Thus several successive year-classes that were abundant as 0-group (1984, 1985 and 1986) became poor at age 3 due to high mortality as one- and two-year-olds (Bogstad and Mehl 1992; Jakobsen 1995), and the recruitment predictions failed completely. However, when comparing the whole time series of 0-group indices for cod with the corresponding numbers at age 3 from the catch-at-age analysis, the conclusion is that poor year-classes as

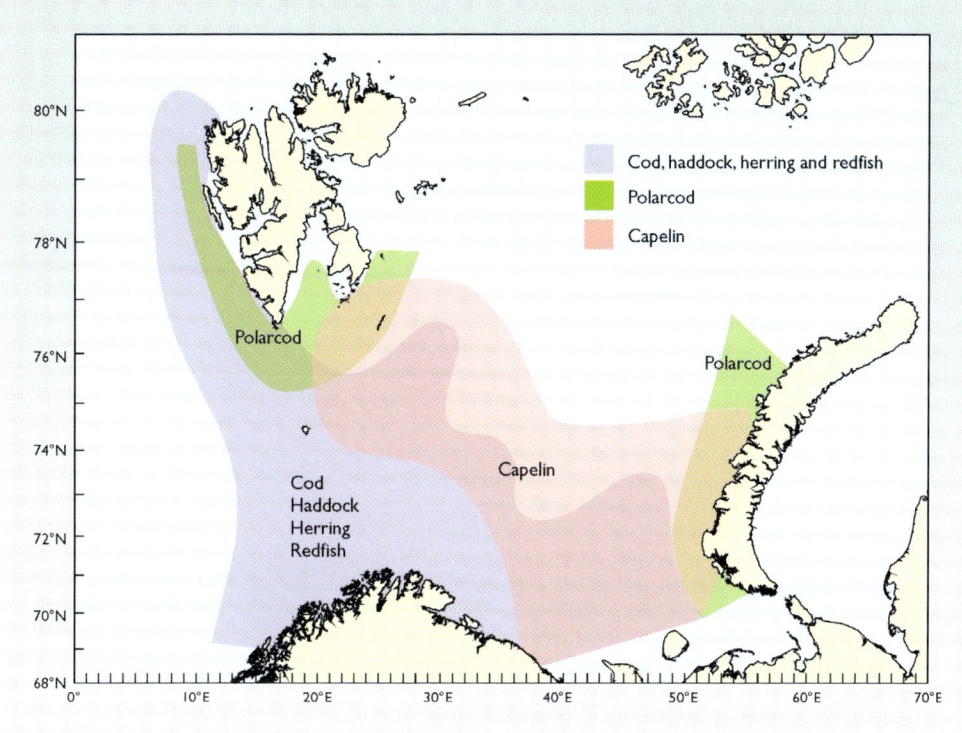

Figure 6.8 Schematic representation of the geographical distribution of the various species in 1985–1998. (From Stensholt and Nakken 2001).

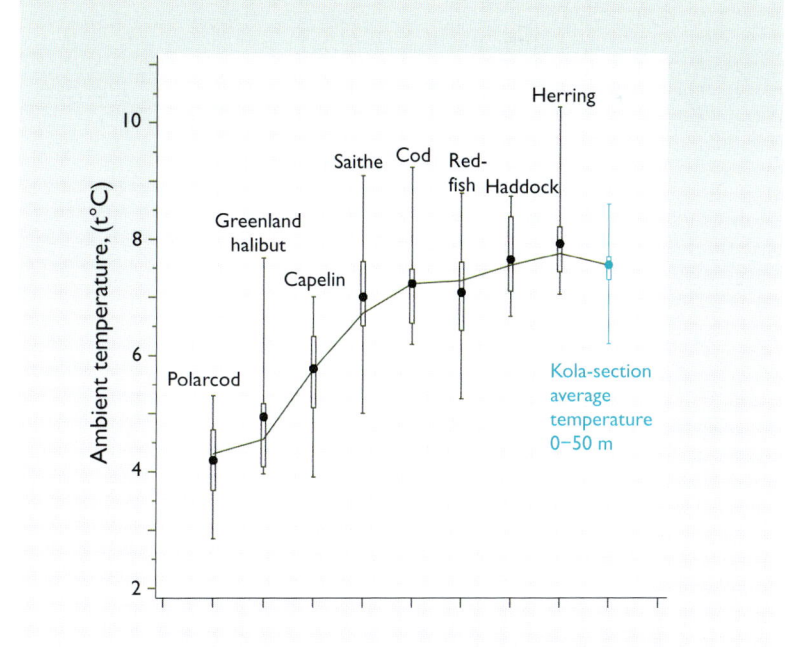

Figure 6.9 Habitat temperature for the 0-group during the annual 0-group surveys in August–September, 1985–1998. The curve joins median values and black dots indicate mean values. Boxes show interquartils and whiskers show full range. (From Stensholt and Nakken 2001).

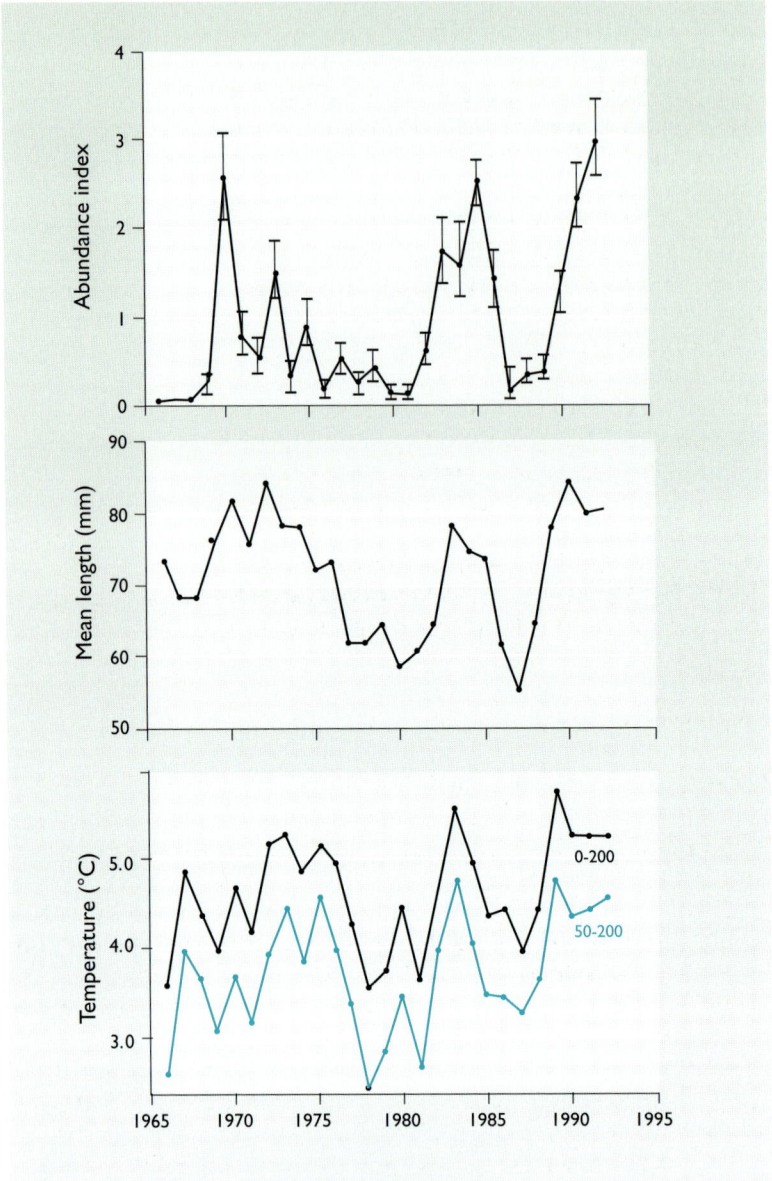

Figure 6.10 Indices (logarithmic) of abundance (upper) and mean lengths (middle) of 0-group cod and mean temperature in the Kola section in August–September. (From Nakken 1994).

0-group remain poor at later stages, while medium and strong year-classes as 0-group may attain a wide range of numbers at age 3 because of differences in mortality during the intervening years. Similar results were arrived at for herring by de Barros and Toresen (1995), who found a large inter-cohort variability in juvenile survival, from very low to around 30 percent, between the 0-group stage and age 3 for the 1983–1993 year-classes. Ottersen and Sundby (1995) explored possible causal relationships between ocean climate variability and recruitment of cod using statistical analysis of sea temperature, wind conditions, spawning stock biomass and year-class strength measured at three different life stages; early juveniles, 0-group and age 3. They found that temperature was the most significant environmental factor related to recruitment success, and also that the spawning stock biomass was almost as important to the formation of year-class strength.

Figure 6.10 shows an important result of the survey; the size and abundance of 0-group cod are synchronized and positively correlated with temperature. Ottersen and Loeng (2000), who investigated this feature for 0-group cod, haddock and herring, concluded that "the synchrony found in year-class strength and early growth of three Barents Sea populations is a result of a mutual response to temperature fluctuations. This further leads us to hypothesize more generally that for stocks at the high latitudes and of the overall range of the species, the environmental signal tends to over-ride density-dependent effects on growth". Their results of similarity in variability patterns among the stocks, and also with temperature, strengthen the conclusions of others (Dragesund 1971; Sætersdal and Loeng 1987; Ottersen 1996) that they are strongly and similarly influenced by large-scale environmental processes. However, the observed covariability between growth and year-class strength at the pelagic 0-group stage may change completely at later stages due to the shifts in habitats at the end of the first year of life.

6.5 Impact, value and justification of 35 years of unbroken pre-recruit studies

For more than 30 years, the Barents Sea 0-group surveys have provided annual information about the distribution and state of the new cohorts of Northeast Arctic cod, Norwegian spring-spawning herring and other important commercial species. During the early period, the accuracy of the survey findings was not high, but as the only ones available based on relevant documented data, they were nevertheless valuable at the time for fisheries management. Over the years, however, survey experience generated ideas for improved methodology, and new technological developments provided advanced hardware and software of great usefulness, with the result that the quality and reliability of the survey results continuously improved. Their value and importance for stock assessments and subsequent management decisions have thus persevered and grown. In addition the survey results have contributed significantly to our understanding of how the environment, including species and stock interactions, affect recruitment and thereby the future fisheries. For these reasons alone, the uninterrupted sequence of annual, comprehensive multi-ship surveys during more than 35 years has clearly been justified.

The development of a survey methodology that focused from the very beginning on the continuous recording of an "echo index" as a measure of abundance, combined with trawl catches for species identification and size determination, set a standard for the conduct of acoustic surveys that has since been applied in many regions of the world. Furthermore, the initially felt need for unbiased quantification of the 0-group echo-recordings led to the invention of the echo-integrator that marked a paradigm shift in fish abundance determination from hydro-acoustic survey data. This paved the way for fast and extensive developments of hardware and software designed specifically for fishery research that have become widely employed, particularly in stock assessments of pelagic fish.

The Barents Sea 0-group survey also initiated a closer cooperation between IMR in Bergen and PINRO in Murmansk. This cooperation was gradually extended to cover many fields of fisheries research and became of mutual benefit for the two institutions as well as for the management of the resources in the region.

REFERENCES

Beltestad, A., Nakken, O., Smedstad, O. 1975. Investigations on diurnal, vertical migration of 0-group fish in the Barents Sea. Fiskeridirektoratets Skrifter, Serie Havundersøkelser, 16: 229–244.

Benko, Y.K., Dragesund, O., Hognestad, P.T., Jones, B.W., Monstad, T., Nizovtsco, G.P., Olsen, S., Seliverstov, A.S. 1970. Distribution and abundance of 0-group fish in the Barents Sea in August–September 1965–1968. ICES Cooperative Research Report, Serie A, No 18: 35–52 and Figures 21–61.

Bogstad, B., Mehl, S. 1992. The Northeast Arctic cod stock's consumption of different prey species 1984–1989. Pp. 59–72 in: Bogstad, B., Tjelmeland, S. (eds.): Interrelations between fish populations in the Barents Sea. Proceedings of the fifth PINRO–IMR symposium, Murmansk, 12–16 August 1991. Institute of Marine Research, Bergen, Norway.

de Barros, P., Toresen, R. 1995. Modelling age-dependent natural mortality of juvenile Norwegian spring-spawning herring (*Clupea harengus* L.) in the Barents Sea. Pp 243–261.

Dragesund, O., Hognestad, P.T. 1960. Småsildundersøkelsene og småsildfisket 1959–60. Fiskets Gang, 46: 703–714.

Dragesund, O., Hognestad, P.T. 1962. Småsild- og feitsildtokt med FF "G.O. Sars" i tiden 30. august til 28. september 1961. Fiskets Gang, 48: 6–11.

Dragesund, O., Olsen, S. 1965. On the possibility of estimating year-class strength by measuring echo-abundance of 0-group fish. Fiskeridirektoratets Skrifter, Serie Havundersøkelser, 13, (8): 48–75.

Dragesund, O. (ed.) 1970. International 0-group fish surveys in the Barents Sea 1965–1968. ICES Cooperation Research Report, Serie A, No 18.

Dragesund, O. 1971. Comparative analysis of year-class strength among fish stocks in the North Atlantic. Fiskeridirektoratets Skrifter, Serie Havundersøkelser, 16: 49–64.

Dragesund, O., Midttun, L., Olsen, S. 1970. Methods for Estimating Distribution and Abundance of 0-group Fish. ICES Cooperative Research Report, Serie A, No 18 (2): 25–34.

Godø, O.R., Valdemarsen, J.W., Engås, A. 1993. Comparison of efficiency of standard and experimental juvenile gadoid sampling trawls. ICES Marine Science Symposia, 196: 196–201.

Gulland, J.A. (ed.) 1964. Contributions to symposium 1963. On the Measurement of Abundance of Fish Stocks. Rapports et Procès Verbaux des Réunions du Conseil International pour l'Exploration de la Mer, Vol. 155.

Haug, A., Nakken, O. 1977. Echo abundance indices of 0-group fish in the Barents Sea 1965-1972. Rapports et Procès Verbeaux des Réunions du Conseil International pour l'Exploration de la Mer, 170: 259–264.

Helle, K., Bogstad, B., Marshall, T., Michalsen, K., Ottersen, G., Pennington, M. 2000. An evaluation of recruitment indices for Arcto-Norwegian cod (*Gadus morhua* L). Fisheries Research 48: 55–67.

Hylen, A. 1997. Acoustic abundance estimate of 0-group Northeast Arctic cod and haddock. ICES CM 1997/B:17: 1–8.

Hylen, A., Korsbrekke, K., Nakken, O., Ona, E. 1995. Comparison of the capture efficiency of 0-group fish in pelagic trawls. Pp. 145–156 in: Hylen, A. (ed.): Precision and relevance of pre-recruit studies for fishery management related to fish stocks in the Barents Sea and adjacent waters. Proceedings of the sixth IMR–PINRO Symposium, Bergen, 14–17 June 1994. Institute of Marine Research, Bergen, Norway.

Hylen, A. (ed.) 1995. Proceedings of the sixth IMR–PINRO symposium, Bergen, 14–17 June 1994. Institute of Marine Research, Bergen, Norway. ISBN 82-7461 039 3.

Hylen, A. (ed.) Precision and Relevance of pre-Recruit studies for Fishery Management related to Fish stocks in the Barents Sea and adjacent waters. Proceedings of the sixth IMR–PINRO Symposium, Bergen, June 1994.

ICES, 1965. Preliminary Report of the joint Soviet-Norwegian investigations in the Barents Sea and adjacent waters September 1965. ICES CM 1965. No 161. (Mimeo.)

ICES, 1966. Preliminary Report of the joint international 0-group fish survey in the Barents Sea and adjacent waters August/September 1966. ICES CM 1966/H:23. (Mimeo.)

ICES, 1967. Preliminary Report of the international 0-group fish survey in the Barents Sea and adjacent waters August/September 1967. ICES CM 1967/H:31. (Mimeo.)

ICES, 1968. Preliminary Report of the 0-group fish survey in the Barents Sea and adjacent waters in August/September 1968. ICES CM 1968/F:33. (Mimeo.)

ICES, 1969. Preliminary Report of the 0-group fish survey in the Barents Sea and adjacent waters in August/September 1969. ICES CM 1969. (Mimeo).

ICES, 2003. Report of the Arctic Fisheries Working Group 2003/ACFM 22, ICES CM.

Jakobsen, T. 1995. Precision in the recruitment estimates and its implications for management of demersal fish stocks. Pp. 285–304 in: Hylen, A. (ed.): Precision and relevance of pre-recruit studies for fishery management related to fish stocks in the Barents Sea and adjacent waters. Proceedings of the sixth IMR–PINRO symposium, Bergen 14–17 June 1994. Institute of Marine Research, Bergen, Norway.

Loeng, H., Bjørke, H., Ottersen, G. 1995. Larval fish growth in the Barents Sea. In: Climate Change and Northern Fish Populations, Beamish, R.T. (editor). Canadian Special Publications of Fisheries and Aquatic Sciences, 121.

Nakken, O. 1994. Causes of trends and fluctuations in the Arcto-Norwegian cod stock. ICES Marine Science Symposia, 198: 212–228.

Nakken, O., Raknes, A. 1996. Corrections of indices of abundance of 0-group fish in the Barents Sea for varying capture efficiency. ICES CM 1996/G: 12. 1–10.

Nakken, O., Hylen, A., Ona. E. 1995. Acoustic estimates of 0-group fish abundance in the Barents Sea and adjacent waters in 1992 and 1993. Pp. 187–198 in: Hylen, A. (ed.): Precision and relevance of pre-recruit studies for fishery management related to fish stocks in the Barents Sea and adjacent waters. Proceedings of the sixth IMR–PINRO symposium, Bergen, 14–17 June 1994, Institute of Marine Research, Bergen, Norway.

Olsen, S. 1960. Rapport om tokt med "G.O. Sars" til Barentshavet i september 1960. Fiskets Gang, 46: 690–692.

Ottersen, G., Loeng, H. 2000. Covariability in early growth and year-class strength of Barents Sea cod, haddock and herring; the environmental link. ICES Journal of Marine Science, 57: 339–348.

Ottersen, G., Sundby, S. 1995. Effects of temperature, wind and spawning stock biomass on recruitment of Arcto-Norwegian cod. Fisheries Oceanography, 4: 278–292.

Randa, K. 1982. Recruitment indices for the Arcto-Norwegian cod for the period 1965–1979 based on the international 0-group survey. ICES CM 1982/G: 53, 22 pp.

Randa, K. 1984. Abundance and distribution of 0-group Arcto-Norwegian cod and haddock 1965–1982. Pp. 192–212 in: Godø, O.R., Tilseth, S. (Eds), Proceedings of the Soviet–Norwegian Symposium on Reproduction and Recruitment of Arctic Cod, Leningrad, 25–30, September 1983, Institute of Marine Research, Bergen.

Sundby, S., Bjørke, H., Soldal, A.V., Olsen, S. 1989. Mortality rates during the early life stages and year-class strength of Northeast Arctic cod (*Gadus morhua* L.) Rapports et Procès Verbaux des Réunions du Conseil International pour l'Exploration de la Mer, 191: 351–358.

Sætersdal, G., Loeng, H. 1987. Ecological adaptation of reproduction in Northeast Arctic cod. Fisheries Research, 5: 253–270.

Stensholt, B.K., Nakken, O. 2001. Environmental factors, spatial density and size distributions of 0-group fish. In: Spatial Processes and Management of Marine Populations, Alaska Sea Grant College Program. AK-SG-01-02, 2001. 395–413.

Toresen, R. 1985. Recruitment indices of Norwegian spring-spawning herring for the period 1965–1984 based on the international 0-group fish surveys. ICES CM 1985/H:54, 1–10.

Toresen, R., Gjøsæter, H., de Barros, P. 1998. The acoustic method as used in the abundance estimation of capelin (*Mallotus villosus*, Muller) and herring (*Clupea harengus*, Linné) in the Barents Sea. Fisheries Research 34 (1998): 27–37.

Valdemarsen, J.W., Misund, O. 1995. Trawl designs and techniques used by Norwegian research vessels to sample fish in the pelagic zone. Pp. 135–144 in: Hylen, A. (ed.): Precision and relevance of pre-recruit studies for fishery management related to fish stocks in the Barents Sea and adjacent waters. Proceedings of the sixth IMR–PINRO Symposium, Bergen, 14–17 June 1994. Institute of Marine Research, Bergen, Norway.

CHAPTER 7

Acoustics in fisheries science in Norway

Odd Nakken

7.1 Introduction

Echo-sounders (and sonars) for seabed detection and mapping came into use as a result of research motivated by the First World War. Two decades later, Sverdrup *et al.* (1942) summarized the benefits of the new and efficient tool as follows: "This new method has in a few years completely altered our concept of the topography of the ocean bottom." During the 1920s, skippers and scientists using echo-sounders for seabed detection noticed that the instrument also recorded echoes from targets in the water column. They assumed these targets were from schools of fish, and the development and use of echo-sounders as fish finders began during the 1930s. At the outbreak of the Second World War, only a few large fishing vessels were equipped with acoustic instruments. Submarine warfare research during the war greatly advanced underwater acoustic technology, and during the two first decades after the war, echo-sounders and sonar came into widespread use in commercial fisheries and fisheries science. In an article compiled by FAO in 1952, we find the following statement (Cushing *et al.* 1952): "Although the fisherman has also benefited from other recent advances in science and engineering, it is safe to say that of all the tools of his trade acquired during this period, the common use of the echo-sounder for detecting fish must be considered the outstanding development of commercial fisheries. Electronic fish detection is already stimulating the development of new fishing gear, such as the midwater trawls, and is taking much of the guess-work out of fishing with the old and established gear in both the pelagic and demersal fisheries."

In Norway, both scientists and fishermen were early aware of the potentials of echo-sounders. Fishing vessels from the Møre area were equipped with such instruments in 1932, and captain Thor Iversen at the Directorate of Fisheries used an echo-sounder to map the seabed off East-Greenland in 1933 (Johansen 1989; Iversen 1936). In 1934, the fishing skipper Reinert Bokn demonstrated the advantage of the instrument for fish finding in the purse seine fishery for sprat (Anon. 1934). A year later, Oscar Sund made his famous echo recordings of spawning concentrations of cod in Lofoten (Figure 7.1) with the RV "Johan Hjort". Sund was immediately aware of the usefulness and potential of the echo-sounder both in fisheries and marine science because of the information about the spatial distribution of fish provided by the instrument that offered

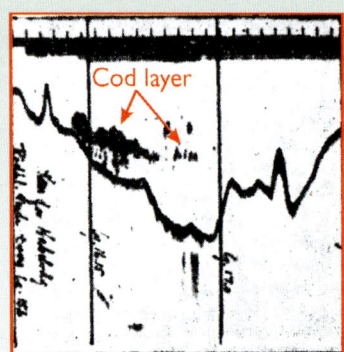 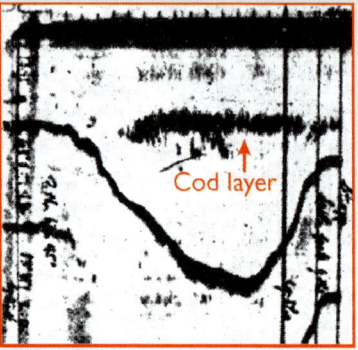

Figure 7.1
Echo-sounder recordings of spawning cod (skrei) in Lofoten in March 1935 (Sund 1935).

the possibility of studying fish behaviour and migration as well as abundance estimation. His observations and analysis were the beginning of a methodological development in Norwegian fisheries science, a development that was interrupted by the Second World War, but continued and expanded after the war through cooperation between marine scientists and engineers. This cooperation kept Norway at the forefront of developments in fisheries and fisheries science acoustics (Schwach 2004; Sogner 1997).

This chapter offers a brief review of how the use of acoustic instruments developed in fisheries science in Norway, from a simple "present or absent" fish indicator in the 1930s to tools providing quantitative information on species, size, abundance and movements of fish and other marine organisms. Readers who wish more detailed information about these developments are referred to earlier reviews (Cushing *et al.* 1952; Dragesund and Midttun 1966; Aglen 1994; Ona 1994; Misund 1997; Fernandez *et al.* 2002; Schwach 2004).

7.2 Echo surveys of cod and herring: the early years

Before 1950, Norwegian fisheries for cod and herring were largely seasonal and coastal, and were based on the concentrations of fish aggregating in coastal waters to spawn in late winter and early spring. The appearance of fish on the spawning grounds varied from one year to the next both temporarily and geographically, and the continuously updated picture of the underwater situation provided by the echo-sounder, enabled scientists to locate and advise the fishing fleet about fishable concentrations. From 1935 until the Second World War, Oscar Sund and his colleagues employed echo-sounders regularly in cod and herring studies, mapping distributions, relating them to environmental conditions and informing the fishing fleets about the results (Sund 1939 and 1943; Runnstrøm 1937 and 1941; Eggvin 1938). The promising immediate economic benefit of such investigations, i.e. advice on where and when to fish, made them attractive to industry and scientists. In an article in Nature (Sund 1935) describing the very first echo-sounder recordings of spawning cod, Sund writes: "It is interesting to note that this spawning concentration of cod had apparently no relation to the bottom. This was well known before, but no one could have imagined the fish to be limited to such a sharply defined layer of only 10–12 metres in thickness, extending widely above deep water and shallow, always at the same distance from the water surface. This distance was 72 metres at the first encounter with the spawning shoals (11 March) and 50 metres at the conclusion of this investigation 5 April. At the same time, the temperature of the "fish" water layer had fallen from 6.0–6.5 °C to about 3.0 °C". Although the instrument was continuously in use in all cod investigations with "Johan Hjort" before the Second World War, it was only in Lofoten during the spawning season that the echo-sounder recorded significant concentrations of cod.

After some preliminary trials of echo surveys of herring in 1936, the echo-sounder was widely used during the herring investigations in 1937. On 7 February, when returning to the coast after completing a hydrographic section across the Norwegian trench, "Johan Hjort" recorded a school of herring "over a distance of 20 nautical miles. The school had a width of 2 nautical miles and a vertical thickness of about 30 metres" (Runnstrøm 1941). Runnstrøm estimated the biomass of the school to be 1.5 million tonnes, assuming a density of two fish per cubic metre, a figure he had arrived at some days earlier when an experienced master fisherman had used his manual "hand sounder" (see Figure 3.2) to assess the size of a smaller school that first had been measured by the echo-sounder on board "Johan Hjort". In his work, Runnstrøm used echo-sounder observations in studies of dial behaviour, abundance and distribution in relation to hydrographic conditions (Figures 7.2 and 7.3). He explained how and why the availability of herring to the fisheries varied from year to year independent of the relative strength of the spawning stock. In years when the coastal waters were cold and extended to great depths, the herring would spawn more deeply, out of reach of many of the gears in use at the time. Eggvin (1940) described

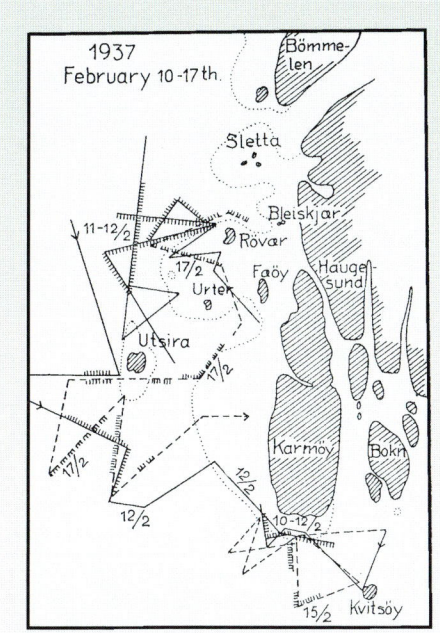

Figure 7.2
Course lines with markings of echo-sounder recordings of spawning herring off south-western Norway in February 1937 (Runnstrøm 1941).

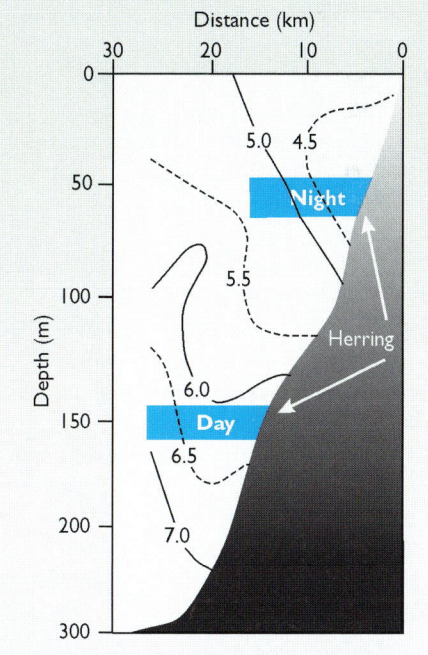

Figure 7.3
Distribution of temperature and echo recordings of herring at a spawning ground in south-western Norway in February 1937 (Redrawn from Runnstrøm 1941).

Gudmund Vestnes preparing for calibration of acoustic instruments. (Photo: H.P. Knudsen).

how an outbreak of cold water and its propagation from the Skagerrak along the coast of southern and western Norway made the herring disappear from the traditional spawning grounds, probably towards deeper waters.

By the outbreak of the war, Norwegian fisheries scientists had been using the echo-sounder regularly for several years and had demonstrated the usefulness of the instrument both for the purpose of observing and understanding fish behaviour, and for advising fishermen on where to fish in order to improve their catches.

The promising practical and scientific experiences with the use of echo-sounders, i.e. the ability to map the spatial distribution of fish and thus obtain information on just where to fish, encouraged Norwegian scientists and engineers to put considerable efforts into developing acoustic fish finders after the war.

Horizontal-ranging sonar was widely used for submarine detection during the war. Sonar operators had observed that echoes from other targets than submarines were recorded, and they assumed that many of these "false echoes" were from fish (herring) schools (Dragesund and Midttun 1966). Immediately after the war, the Institute of Marine Research started trials of sonar for detecting herring schools. During the winter fishery in 1946, a Norwegian Navy vessel, "Eglantine", operated a sonar unit on the herring fishing grounds (Gerhardsen 1946; Lea 1947). The scientist in charge was Einar Lea, the leader of the herring investigations, and the engineer responsible for the operation of the equipment was Thorvald Gerhardsen, who later became the managing director of Simrad. In his report, Lea (1947) identified two research tasks for which the new instrument could be very useful:

- A scouting service in early winter to advise fishermen on when and where the pre-spawning herring would appear at the coast, combined with oceanographic studies aimed at understanding the large shifts in spawning migrations between years.
- An acoustic survey in summer and autumn covering most of the Norwegian Sea and the western parts of the Barents Sea to map the distributions of mature and immature herring in the feeding season.

Trials continued during the 1947 and 1948 herring fishing seasons, with Finn Devold as cruise leader. The outcome of these experiments was that a new sonar system, specially designed for fish finding, was installed on board IMR's new research vessel "G.O. Sars" in 1950. A new echo-sounder built by Simrad (established in 1947), was also tested by Devold and his colleagues on board the purse-seiner "Vartdal" during a cruise in the Norwegian Sea in summer 1949. In his report, Devold concluded that the instrument performed as well as instruments of foreign origin, and the new sounder became a sales success (Sogner 1997). Thus, in only four or five years, close cooperation between fishery scientists and engineers had resulted in Norwegian acoustic systems (echo-sounder and sonar) of high quality. This was the start of a fruitful cooperation between IMR and Simrad that has been maintained throughout the years, a cooperation in which Gudmund Vestnes at IMR played a key role for 40 years, and from which both institutions have benefited.

The new research vessel with her instrumentation became a great success: "It soon became apparent that the "G.O. Sars" was worthy not only of her name but of the four million Norwegian kroner that were spent on her," wrote Rollefsen (1966). During the summer of 1950 she mapped the herring distribution in the Norwegian Sea, and in winter 1950–1951 she traced the spawning migration of herring towards the Norwegian coast (Figure 7.4). Daily bulletins

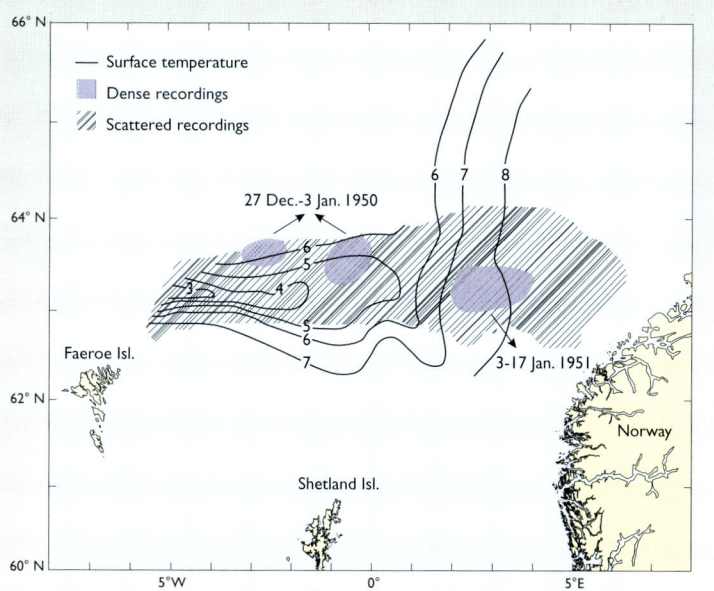

Figure 7.4 Distribution of surface temperature (°C) and sonar observations of herring migrating towards the Norwegian coast in winter 1950–1951 (Redrawn from Devold 1952).

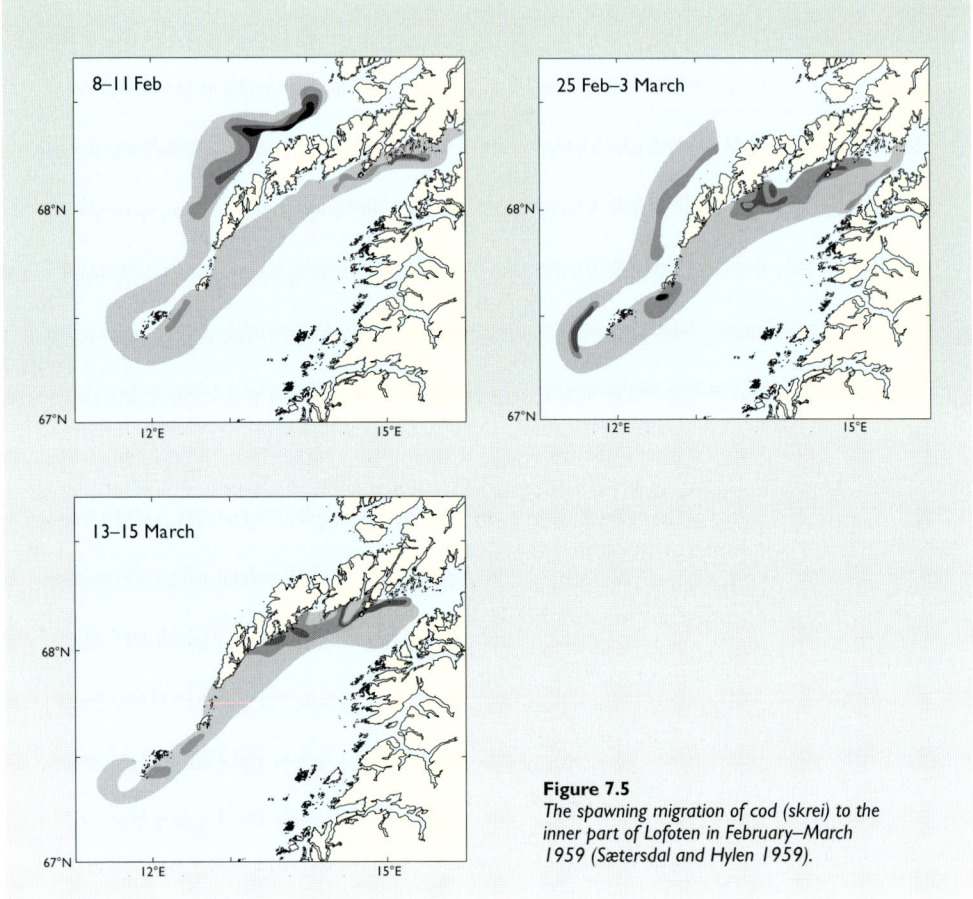

Figure 7.5
The spawning migration of cod (skrei) to the inner part of Lofoten in February–March 1959 (Sætersdal and Hylen 1959).

on the position, numbers, size and depth of schools were broadcast to the fishing fleet, and both the service and the cruise leader and head of the herring investigations, Finn Devold, became extremely popular in the herring fishing communities (see Chapter 4).

A similar service was established for the "skrei" fishery, tracking the spawning migration of Northeast Arctic cod moving southward from the Barents Sea towards the Lofoten region, and charts (Figure 7.5) of the distribution were published in a local Lofoten newspaper so that fishermen were kept updated on the movements of fish to the spawning grounds.

As well as providing information to the fishermen on where to fish, the echo surveys contributed greatly to the knowledge of fish distribution, migrations and behaviour in relation to environmental factors, knowledge that has become useful in the planning and interpretation of the results of stock assessment survey in later years. The development of fish-finding equipment, capture and communication systems in the fishing fleet between 1950 and 1970 (see Chapter 3) rendered largely unnecessary the need to provide direct advice to the fishermen, and the main objectives of the echo surveys shifted towards obtaining estimates of abundance of fish stocks.

7.3 Towards quantification, 1950–1970

By the early 1950s, it was well known that the strength of the echo (the echo amplitude) was seriously affected by the presence or absence of a swim-bladder, and it was soon realized that a number of other factors were also important: instrument characteristics, such as the beam pattern of the transducer and acoustic frequency; depth, recognizing the need for time-varied (depth-varied) gain; and target characteristics, such as the size and shape of individual fish. However, empirical knowledge that would make it possible to quantify the effect of the various factors on echo recordings was limited. During the 1950s and 60s, several methods were developed to transform echo recordings into estimates of fish density (number per unit volume or number per unit area) by counting echo traces, measuring their size on the paper record or measuring the voltage of the echo (see Fernandez *et al.* 2002 for references).

Scientists at the Institute of Marine Research were active in this work and were at the forefront of developments. The observation that skilled purse seine fishermen soon developed an ability to estimate the biomass of herring, sprat and saithe schools rather accurately from echo-sounder and sonar recordings stimulated efforts in acoustic abundance estimation. They were early aware that cod, for instance, were more widely dispersed and pelagically distributed at night than during the day, when the fish were more tightly clustered and stayed closer to the bottom, and they used this difference in diurnal behaviour in practical surveying, performing echo surveys at night and sampling (trawling) the fish during the day. Sætersdal (1955) recognized traces of medium-sized single cod down to 170–180 m, and Midttun and Sætersdal (1957) made the first attempt to arrive at an absolute estimate of abundance. They estimated the number of fish within a limited area in the Barents Sea by counting individual fish traces on the recording paper. The sampling angle of the echo-sounder – i.e. the angle within which the sounder "saw" the fish – was found by counting the number of echoes in each fish trace and comparing the values arrived at with the actual beam pattern of the transducer (Midttun and Vestnes 1977). At times the group (Midttun, Sætersdal and Vestnes) had difficulties in separating echoes from cod from the bottom echo, a problem that became much less when in 1956, in cooperation with Simrad engineers, they succeeded in establishing the so-called "white line" bottom by removing the peak values from each seabed echo so that it appears on the recorder as a thin black line followed by a broader white line (Midttun, Sætersdal and Vestnes 1957). The experience gained during the 1950s were so promising that in 1961/1962, IMR installed a new echo-sounder and sonar system on board the two research vessels "Johan Hjort" and "G.O. Sars" (Dragesund and Midttun 1966; Midttun and Vestnes 1977). This instrument included automatic compensation for transmission loss (one-way and two-way) of sound intensity so that densities and targets at different depths could be compared more reliably, and target strength measurements could be made more easily at sea (Midttun 1966; Midttun and Nakken 1968).

Fish density (and abundance) from echo counting can only be estimated when fish are so well separated from their neighbours that echoes do not overlap. For recordings of scattering layers where two or more fish contribute to each echo, a subjective visual density grading on a scale from 0 (no recording) to 4 (very dense recording) was developed and used until the echo integrator was introduced in 1963 (Dragesund and Olsen 1965, see Chapter 6). That instrument proved so successful that it formed the basis of modern acoustic abundance estimation (Midttun and Nakken 1968; Dragesund 1970; McLennan 1990). The first version of the instrument integrated echo voltage over a depth range and distance (usually one nautical mile). However, soon it was realized that the echo intensity (i.e. voltage squared) should be integrated, and measurements made by Russian scientists (Scherbino and Truskanov 1966) on herring schools showed empirically that squared echo voltage (echo intensity) is proportional to fish density. "Squared voltage" echo integrators became available from Simrad in 1969 (Bodholt 1969) and were installed on board the new RV "G.O. Sars" in 1970. That vessel was designed and equipped for acoustic measurements, abundance estimation in particular, based on the methodology which then had been successfully employed in the 0-group surveys (Chapter 6). In addition to her ability to make acoustics measurements, her capacity for fish capture and thus for identification and size determination of the acoustic recordings was excellent.

Experiments carried out in the 1960s, greatly improved our knowledge of the back-scattering properties of fish, i.e. their ability to reflect sound (see Fernandez et al. 2002 for references). The results obtained by Midttun and Hoff (1962) are of particular interest and importance for both acoustic abundance estimation and size determination, as they demonstrated that even small changes in fish behaviour (tilt angle) would cause large changes in target strength (scattering cross-section), and that cod and saithe were substantially different in this respect (Figure 7.6). Midttun and Hoff (1962) attributed this to the different shape of the swim bladder in the two species, the saithe having a more elongated swim bladder than the cod. As results from these and other fully controlled experiments gradually became available, the complexity of the relationship between the echo and fish species, size, aspect angle and density was realized, and the demand for better controlled measurements grew. A major step forward was the methodology for estimating the density distribution of the target strength of individual targets (Craig and Forbes 1969). It provided true estimates of numbers per unit volume (or area) for each target-size group by removing the effects of the beam pattern of the transducer from the echoes.

The significant developments in the use of acoustic methods in fisheries science during the 1960s, persuaded ICES to collaborate with FAO in organizing

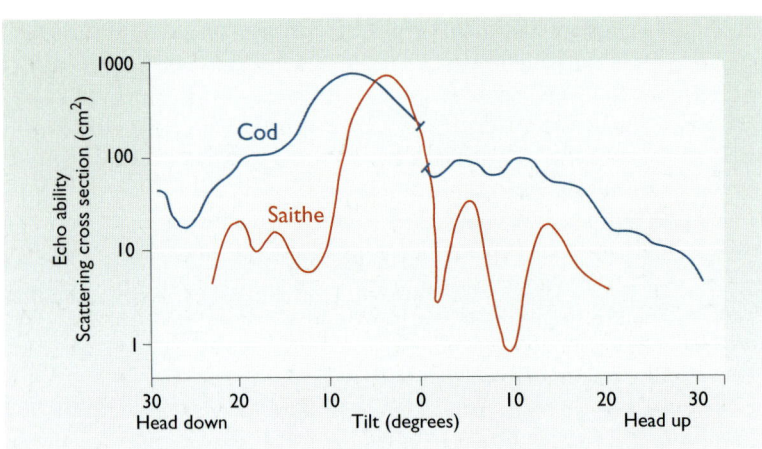

Figure 7.6
Dorsal aspect target strength patterns of a cod (46 cm in length) and a saithe (46 cm in length) at 38 kHz (Redrawn from Midttun and Hoff 1962).

a training course for fisheries scientists in Svolvær (Lofoten) in northern Norway in March 1969. The course produced a number of preliminary manuals, culminating in the work of Forbes and Nakken (1972). In reviewing the outcome of the training course, the ICES' Gear and Behaviour Committee made two recommendations: 1) that an acoustic group be set up within the Committee; and 2) that steps be taken to organize a symposium on "Acoustic Methods in Fisheries Research". The former would not come about until later; the second went ahead and led to the symposium being held in Bergen in 1973 (Margetts 1977).

7.4 Fish stock monitoring – 1970s

By 1970, the increased exploitation of pelagic fish in the Northeast Atlantic in previous years, largely caused by the development in purse seine fishing (see Chapter 3), had brought the stock of Norwegian spring-spawning herring to collapse, reduced the North Sea stocks of herring and mackerel substantially, and increased fishing pressure on the stock of Barents Sea capelin. At the Institute of Marine Research it was decided to use the new and well equipped "G.O. Sars" to obtain acoustic estimates of capelin and blue whiting; capelin because the fishery had increased rapidly throughout the 1960s, and because other and more traditional methods of estimating abundance were difficult to apply on the stock and blue whiting because Scottish and Russian studies had indicated that the stock was large and that yields could be greatly increased. Previous studies had shown that both these stocks had distribution patterns that favoured acoustic abundance estimation, i.e. the fish were aggregated in pelagic scattering layers and to a large extent unmixed with other species.

Figure 7.7 shows the distribution of capelin in late summer and early autumn in the Barents Sea in 1971 and 1972. The difference in geographical extent is mainly caused by the coverage in 1972 being 5–6 weeks earlier than in 1971. Acoustic estimates of abundance came to 6.7 (in 1971) and 9.1 (in 1972) million tonnes of capelin. These "absolute" values were arrived at using a conversion factor for the echo integration values based on paper counts of individual fish traces, a method suggested by Midttun and Nakken (1971) (Figure 7.8), and were explained and discussed in a paper at the acoustic symposium in Bergen in 1973 (Midttun and Nakken 1977), which also presented the results obtained in 1972 and 1973 for blue whiting. Applying a similar calibration technique for blue whiting as for capelin, the four coverages (two in 1972 and two in 1973) resulted in abundance estimates of between 2.7 and 10.4 million tonnes (Midttun and Nakken 1977). The main reasons for estimating the conversion factor using fish densities obtained from paper counts was the uncertainty involved in calibration of the integration system (see below) as well as the lack of target-strength measurements of both species.

The promising results arrived at in the early 1970s for capelin and blue whiting, led to IMR carrying out annual acoustic abundance surveys for both these species as well as for many other stocks. For capelin, the first catch quota (TAC) was introduced for the winter fishery of spawners in 1974, on the basis of the survey results from autumn 1973. For blue whiting, which were observed to be distributed in dense scattering layers during spawning (Figure 7.9), a trial fishery with pelagic trawls started in 1972, ultimately leading to an important Norwegian fishery on this stock (see Chapter 3). Gjøsæter *et al.* (1998) have

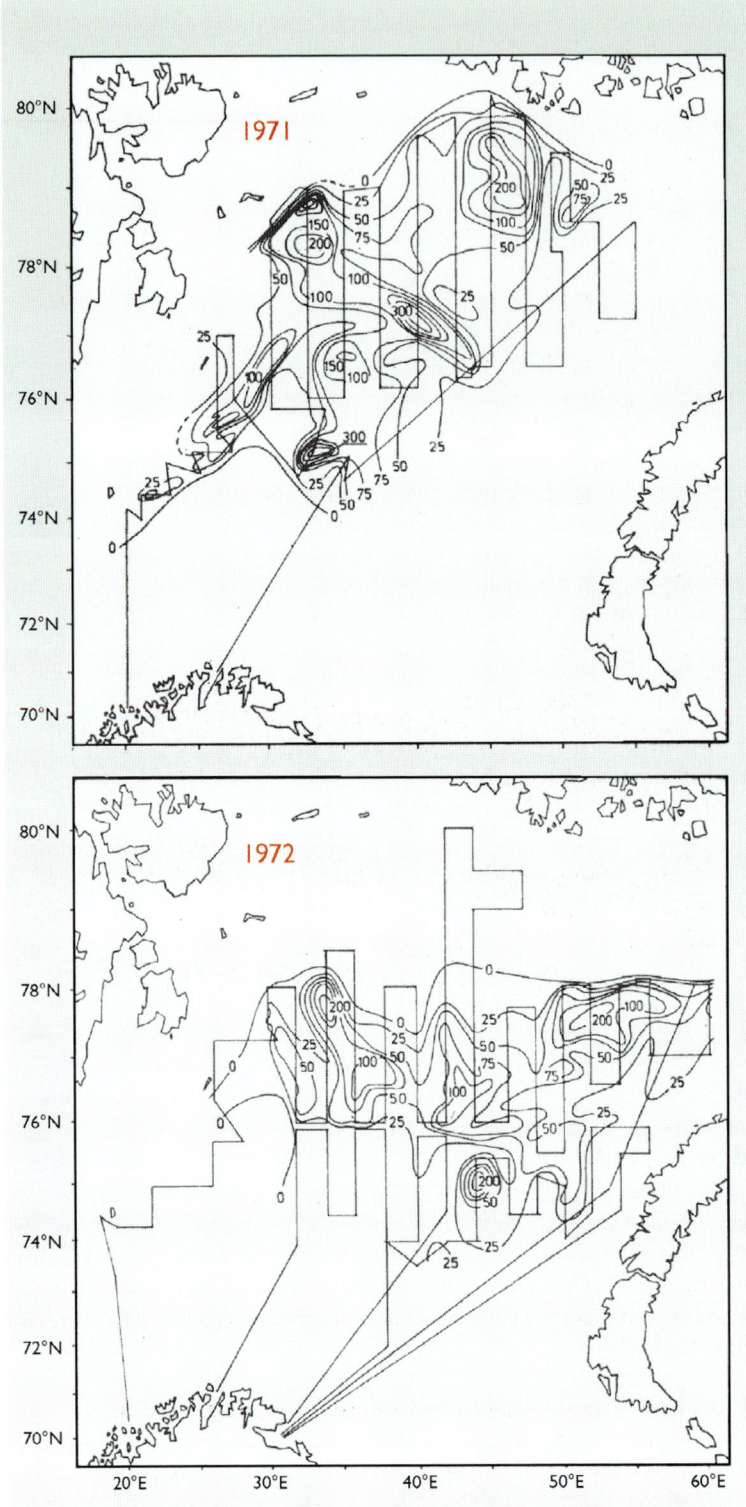

Figure 7.7
The distribution of echo integrator values (echo densities) of capelin in September 1971 and August 1972 (Midttun and Nakken 1977).

Figure 7.8
Corresponding values of echo integrator observations and capelin densities calculated from counts on the echo-sounder paper record (Midttun and Nakken 1977).

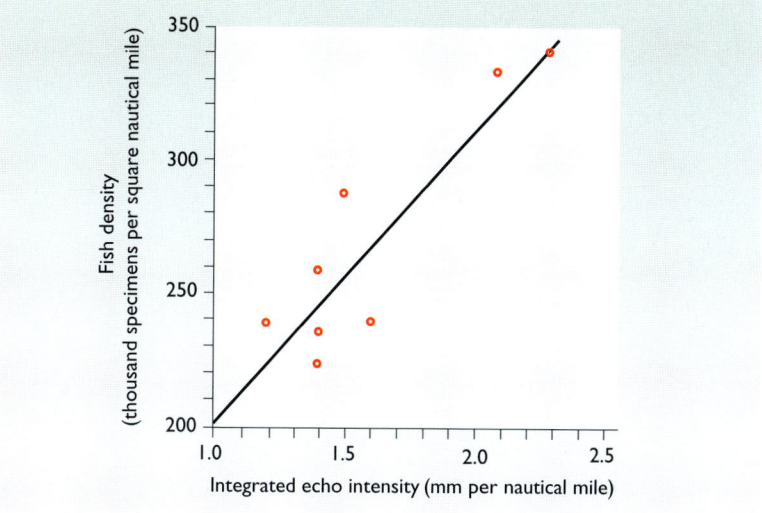

Figure 7.9
Echo-sounder recording of a dense layer of blue whiting at 300–400 m depth in March 1972. Distance between vertical lines is one nautical mile (Midttun and Vestnes 1977).

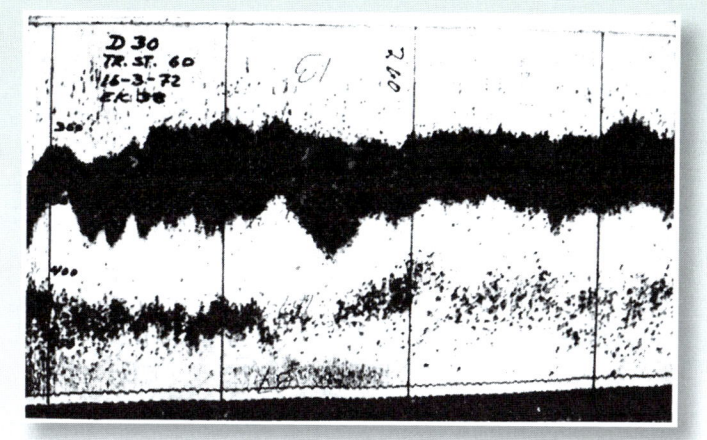

written a comprehensive review and synthesis of the results of the capelin acoustic surveys, 1972–1997, and Monstad (2004) reviewed the information gained from the acoustic surveys on blue whiting. Results for Northeast Arctic cod are discussed in Chapter 5.

Throughout the 1970s, the stock of Norwegian spring-spawning herring was extremely low and was found at a few sites in inshore waters, a distribution pattern that did not favour acoustic abundance estimation. In the 1980s, when the herring stock began to recover, annual acoustic estimates of abundance became an important source of information in the assessment for that stock as well (Chapter 4).

7.5 Increasing understanding and improving methodology

In order to utilize the large quantity of acoustic field data that became available thanks to the echo integration technique developed in the 1960s, more insight was needed into the scattering properties of fish and other marine organisms, including an understanding of how the characteristics and settings of the instruments (frequency, beam pattern, gain and threshold levels) influenced the echoes. Throughout the 1970s and 80s, scientists at the Institute of Marine Research carried out a series of fully controlled experiments in order to obtain more knowledge of the relationship between the echo and fish species, size, orientation relative to the acoustic beam, and density.

Target strength in relation to fish species and size
Nakken and Olsen (1977) measured more than 300 fish of 17 different species at two acoustic frequencies (38 and 120 kHz) in summer 1971, and some years later, Dalen *et al.* (1976) carried out target strength measurements of capelin and 0-group fish.

The results of Nakken and Olsen were presented at the Bergen acoustic symposium in 1973 and were the subject of widespread attention in the following years. Figure 7.10 shows the principal findings for cod. When they combined the tilt angle distribution of cod observed from field photographs (Olsen 1971) with the results of controlled experiments, they found that the mean target strengths to be expected in field observations for a given fish size were 7–8 dB lower than the maximum values observed. Similar results were also obtained for herring when the tilt angle distributions observed by Beltestad (1973) were employed. The measurements made by Nakken and Olsen (1977) were further analyzed by Kenneth Foote in a series of papers in which he established relationships between the average target strength and fish length for a number of species (see Foote 1980 for references). The importance of the size and shape of the swim-bladder for acoustic scattering had long been realized (Midttun and Hoff 1962) and was evident from the measurements made by Nakken and Olsen (1977) when comparing the results for mackerel (no swim bladder) with those for swim bladder fish (Foote 1980). Comprehensive measurements in the early 1980s (Ona 1982) of swim-bladder size and shape offered the possibility of predicting fish target strengths (Foote and Ona 1985; Foote 1985). However, various physiological factors such as stomach content and the seasonal fat cycle and gonad development cause variations in swim-bladder size and shape, and thus natural variations in fish target strength (Ona 1990).

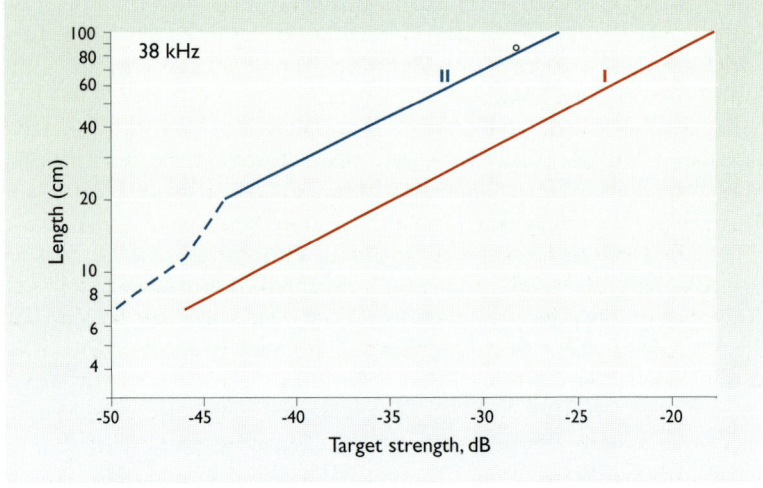

Figure 7.10
Dorsal aspect target strength – length relationship for cod at 38 kHz.
I) Observed maximum target strength, II) Expected target strength when observing at sea, o) Observation at sea (Nakken and Olsen 1977).

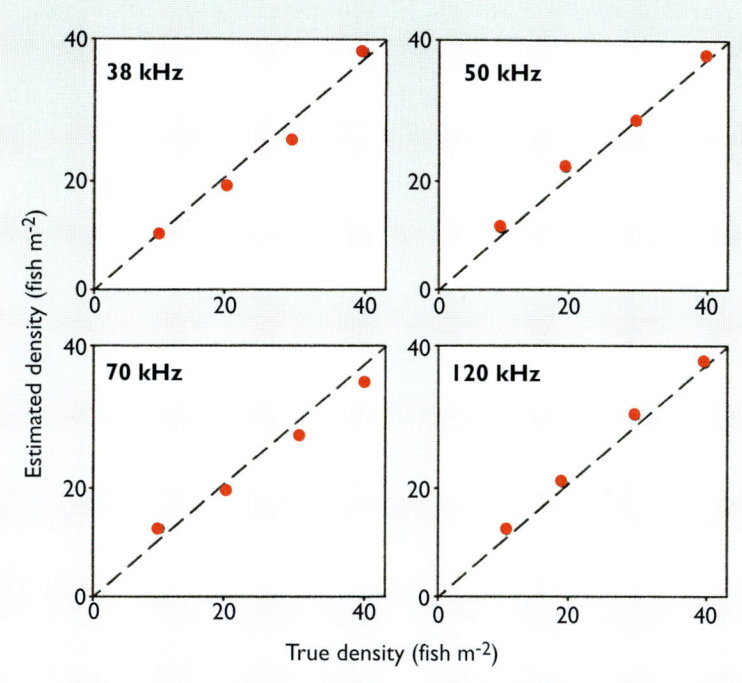

Figure 7.11
Comparison of estimated and true densities of caged fish at four frequencies (Foote 1983).

Echo intensity in relation to fish density
Sometimes the bottom echo below a dense school of fish is reduced compared with its level outside the school, a phenomenon often observed when dense herring schools were recorded. This is often referred to as acoustic shadowing; i.e. the upper parts of the school reduce the sound intensity that reaches the lower parts and the seabed.

In order to clarify the fish density at which acoustic shadowing occurs so that the proportionality with echo intensity breaks down, Røttingen (1976) carried out controlled experiments with several fish species in 1973. His measurements indicated that linearity was maintained up to very high fish densities. The linearity principle, i.e. that integrated echo intensity is proportional to fish density, which is the basis for the use of echo integration in abundance estimation, is not self evident, and discussions continued for years in ICES and other fora about its validity. In the early 1980s, Foote (1983) performed a definitive test of the relationship between echo intensity and fish density. He measured the echo energy from caged free-swimming fish at several frequencies and observed the behaviour of the fish by photography, which enabled him to correct the echo energy for the tilt angle distributions observed. His estimated and true densities agreed within the bonds of experimental error, thus demonstrating the applicability of the linearity principle (Figure 7.11). However, at very high fish densities shadowing does occur, so that acoustic density estimates are biased downward as shown for dense scattering layers and herring schools (Aglen 1994; Olsen 1986 and 1990; Toresen 1991). In recent years, estimation and compensation models for shadowing effects in dense fish aggregations have been developed (Zhao and Ona 2002).

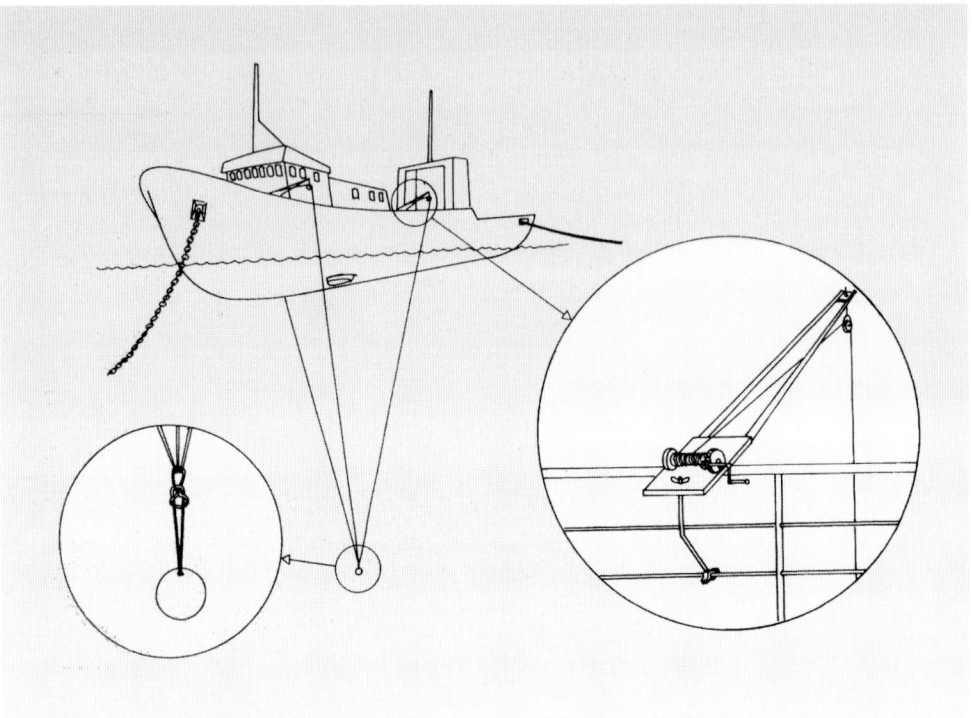

Figure 7.12
Rigging of a research vessel for calibration of the echo-sounder and echo integration system.

Advances in calibration and instrumentation

An essential advance in acoustic abundance estimation and sizing was the progress made in the early 1980s on calibrating acoustic instruments (Foote *et al.* 1983, 1987). Earlier calibrations, based on measurements with hydrophones, were tedious and often produced highly variable results. The new calibration methodology, which used spherical standard targets, optimized to the specific frequency and placed in easy manoeuvrable systems (Figure 7.12), yielded results with high accuracy. Another major step forward in these years was the development of the split-beam echo-sounder, which made it possible to determine the location of the target in the acoustic beam. The basic development work was done by laboratories in Seattle, USA (Ehrenberg 1983). However, by 1984, Simrad made the first split-beam echo-sounder commercially available, an instrument that became a great success for measuring target strength and sizing fish at sea (Foote *et al.* 1986).

A major problem in acoustic measurements at sea when using hull-mounted transducers is the weakening of the transmitted signal caused by the layer of air bubbles that forms below the vessel in bad weather (Dalen and Løvik 1981). The obvious solution to the problem is to tow the transducer at a depth below the bubble curtain. However, in a surveying situation when trawling for fish sampling is often carried out, using a towed body is inconvenient and time-consuming, and since about 1990, IMR has used protruding transducer platforms on its research vessels. This has significantly reduced the problem of air blocking of the echo sounder signals (Ona 1994).

The introduction of digital signal processing in echo-sounders, sonars and echo integrators greatly improved the efficiency of these instruments for fish behaviour studies (Bodholt and Olsen 1977) and density estimations (Blindheim *et al.* 1982). A major improvement for IMR's work was the development of the Bergen Echo Integrator (BEI), a man-machine interface system that made allocations of echo integrals (echo densities) to species or groups of scatterers considerably more easy and reliable (Knudsen 1990; Foote *et al.* 1991). Figure 7.13 plots mean values of echo density in four areas for six species on the basis of allocations by two teams of observers working independently with the same data sets. Cod was the target species for the survey. The comparison appears to be quite good, as can be seen from the scatter of points around the 1/1 line, except for herring in one of the areas.

Survey grid density

The accuracy of an acoustic estimate of abundance obviously depends on the patterns and patchiness of fish distribution and the density of cruise tracks, and in the 1980s, the design of survey grids was widely discussed (see Simmonds *et al.* 1992 for references). Aglen (1983, 1989a and b) investigated the accuracy of surveys made in different circumstances; fjords and large areas of open sea, and tropical and temperate waters. He showed that in order to obtain a given level of accuracy in a narrow fjord, a denser survey grid is required when the fish are lying in schools (during the day) compared with scattered distributions (at night). Such differences were not observed for surveys in the open sea. In the fjord some 5–20 schools were typically recorded, while in the open sea, several hundred schools were recorded, thus evening out small-scale variations. Aglen also established a relationship between the coefficient of variation and the degree of coverage, i.e. the total length of cruise track divided by the square root of the area to be covered, which became very useful for survey planning.

Abundance estimation and fish behaviour

An important part of IMR's work in the 1980s and 1990s on acoustic abundance estimation was on identification and quantification, and quantification of errors as well as on methodology to reduce errors. The progress of this work benefited

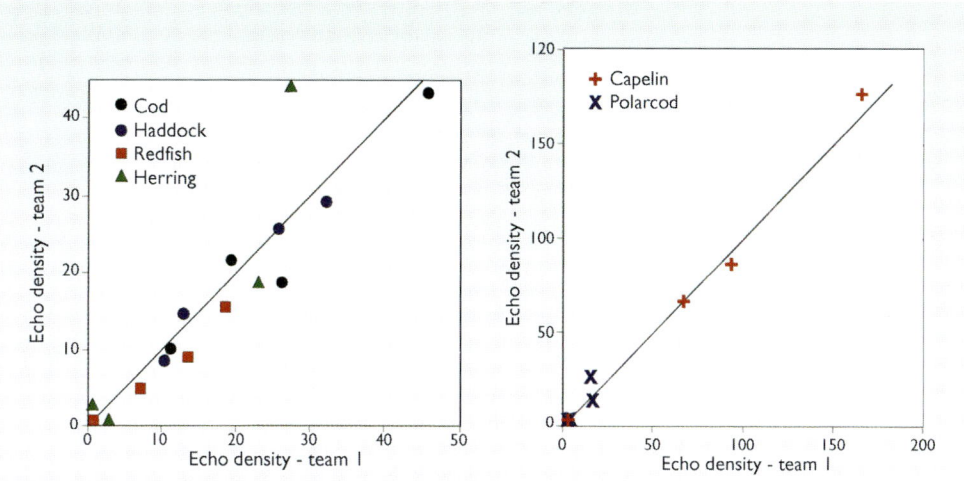

Figure 7.13
Comparison of the results of two teams which allocated echo integration values to fish species. (Korsbrekke and Misund 1993) (Average values for four different areas).

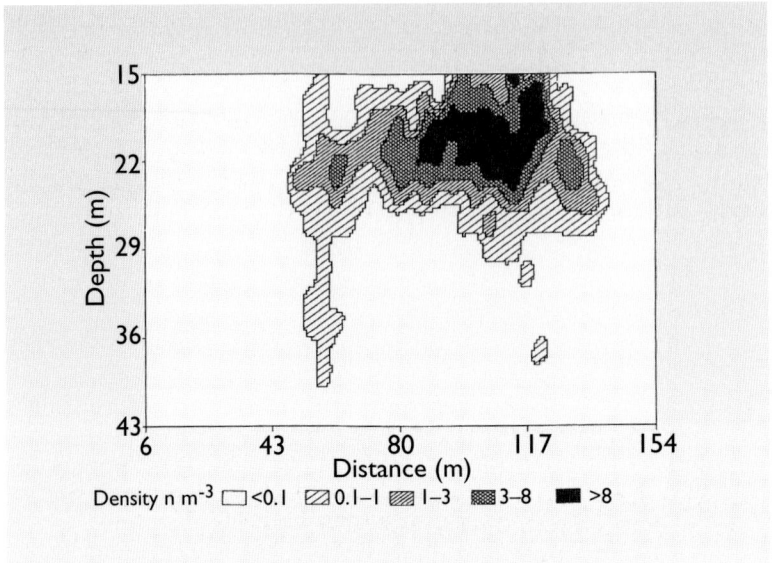

Figure 7.14
Packing density (numbers per cubic metre) structure of a single herring school (Misund 1993).

greatly from cooperation in the Working Group of Fisheries Acoustic Science and Technology (WG FAST) under ICES that met for the first time in 1982 under the chairmanship of Kjell Olsen. Aglen (1994), who reviewed and summarized the available knowledge, considered errors due to fish avoiding the surveying vessel horizontally and vertically to be the most serious, both because such errors may became large and because it is difficult to detect avoidance properly. Observations, particularly on herring, showed that schools and scattering layers near the surface avoided the vessel horizontally (see Chapter 8) while deeper fish also dived as vessels passed, a pattern of behaviour that was modelled by Olsen et al. (1983) by taking into account the spatial distribution of the noise from the survey vessel. Avoidance behaviour could thus both thin fish density along the track of the vessel and reduce target strength due to changes in tilt angle and depth (compression of swim bladder). These effects would lead to underestimation of the actual densities along the track using the echo integration technique (Vabø 1998). Comprehensive field measurements in the 1980s and 90s greatly improved our understanding of fish behaviour patterns, both natural (see Nøttestad et al. 2002 for references) and vessel-induced, and thus our ability to improve acoustic abundance estimation by counting and measuring schools of fish using sonar (see Chapter 8). Sonar counting and size measuring of schools were well established by 1970 (see Forbes and Nakken 1972 for references). However, improvements in technology during the subsequent decades allowed horizontal and vertical school structures and fish movements to be surveyed (Misund 1993; Misund et al. 1993). Figure 7.14 shows the density structure of a herring school.

Discrimination between different types of recordings
Scientists and fishermen were early aware that some of the echo recordings obtained were from planktonic organisms, and in practical surveys, the differences in appearance on the paper record or screen were used to decide whether plankton or fish were being recorded. The use of a range of acoustic frequencies

on board the vessel eased this type of work because of great differences in frequency response of plankton and fish recordings. During the 1990s, methods were developed for quick and easy comparison of differences in frequency response, and these were incorporated in the Bergen Echo Integrator post-processing system (Korneliussen 2002). In addition to making the discrimination between fish and plankton more easy and reliable, the technique has also proved useful for separating recordings of mackerel from those of other fish species.

REFERENCES

Aglen, A. 1983. Random errors of acoustic fish abundance estimates in relation to the survey grid density applied. FAO, Fisheries Report, 300: 293–298.

Aglen, A. 1989a. Empirical results on precision effort relationships for acoustic surveys. ICES CM 1989/B: 30.

Aglen, A. 1989b. Reliability of acoustic fish abundance estimates. Dr.scient. thesis, University of Bergen, Bergen, Norway.

Aglen, A. 1994. Sources of Error in Acoustic Estimation of Fish Abundance. In: Fernøe, A., Olsen, S. (eds.): Marine Fish Behaviour in Capture and Abundance Estimation. Fishing News Books, 1994: 221 pp.

Anon. 1934. Forsøk med ekkolodd i brislingfisket i 1934 [Echo sounding experiments during the sprat fishing season in 1934]. Tidsskrift for Hermetikkindustrien 20: 222–223. (In Norwegian).

Beltestad, A.K. 1974. Beiteatferd, vertikalvandring og stimdannelse hos 0-gruppe sild (*Clupea harengus* L.) i relasjon til lysintensiteten (Feeding behaviour, vertical migration and schooling of 0-group herring in relation to light intensity). Cand.real. thesis, University of Bergen, Bergen, Norway.

Bodholt, H. 1969. Quantitative measurements of scattering layers. SIMRAD Bulletin (3): 1–11.

Bodholt, H., Olsen, K. 1977. Computer-generated display of an underwater situation: applications in fish behaviour stadies. Rapports et Procès-verbaux des Réunions du Conseil International pour l'Exploration de la Mer, 170: 31–35.

Craig, R.E., Forbes, S.T. 1969. A sonar for fish counting. Fiskeridirektoratets Skrifter, Serie Havundersøkelser, 15: 210–219.

Cushing, D.H., Devold, F., Marr, J.C., Kristjonsson, H. 1952. Fisheries Bulletin, FAO, 5 (3–4): 1–27.

Dalen, J., Løvik, A. 1981. The influence of wind-induced bubbles on echo integration surveys. Journal of the Acoustical Society of America, 69: 1653–1659.

Dalen, J., Raknes, A., Røttingen, I. 1976. Target strength measurements and acoustic biomass estimation of capelin and 0-group fish. ICES CM 1976/B: 37.

Devold, F. 1952. A contribution to the study of the migrations of Atlantic-Scandian herring. Rapports et Procès-Verbaux des Réunions du Conseil International pour l'Exploration de la Mer, 131: 103–107.

Devold, F. 1963. The life history of the Atlantic-Scandian herring. Rapports et Procès-Verbaux des Réunions du Conseil International pour l'Exploration de la Mer, 154: 98–108.

Dragesund, O. 1970. Distribution, abundance and mortality of young and adolescent Norwegian spring-spawning herring (*Clupea harengus* Linné) in relation to subsequent year-class strength. Fiskeridirektoratets Skrifter, Serie Havundersøkelser, 15: 451–556.

Dragesund, O., Midttun, L. 1966. The development of acoustic instrumentation in fisheries research and commercial fisheries in Norway. ICES CM 1966/F: 8.

Dragesund, O., Olsen, S. 1965. On the possibility of estimating year-class strength by measuring echo abundance of 0-group fish. Fiskeridirektoratets Skrifter, Serie Havundersøkelser, 13(8): 48–75.

Eggvin, J. 1938. Trekk fra Nord-Norges oceanografi sett i sammenheng med torskefisket. Fiskeridirektoratets Skrifter, Serie Havundersøkelser, 5 (7): 33–46. (In Norwegian with summary in English).

Eggvin, J. 1940. The Movements of a Cold Waterfront. Fiskeridirektoratets Skrifter, Serie Havundersøkelser, 6(5): 151 pp.

Ehrenberg, J.E. 1983. A review of *in situ* target strength estimation techniques. FAO Fisheries Report 300: 85–90.

Fernandez, P.G., Gerlotto, F., Holliday, D.V., Nakken, O., Simmonds, E.J. 2002. Acoustic applications in fisheries science: the ICES contribution. ICES Science Symposia 215: 483–492.

Foote, K.G. 1980. Importance of the swim bladder in acoustic scattering by fish: A comparison of gadoid and mackerel target strengths. Journal of the Acoustical Society of America, 67(6): 2084–2089.

Foote, K.G. 1983. Linearity in fisheries acoustics, with additional theorem. Journal of the Acoustical Society of America, 73: 1932–1940.

Foote, K.G. 1985. Rather high-frequency sound scattering by swim bladder fish. Journal of the Acoustical Society of America, 78: 688–700.

Foote, K.G., Ona, E. 1985. Swim bladder cross sections and acoustic target strengths of 13 pollack and 2 saithe. Fiskeridirektoratets Skrifter, Serie Havundersøkelser, 18: 1–57.

Foote, K.G., Aglen, A., Nakken, O. 1986. Measurement of fish target strength with a split-beam echo sounder. Journal of the Acoustical Society of America, 80: 612–621.

Foote, K.G., Knudsen, H.P., Korneliussen, R.J., Nordbø, P.E., Røang, K. 1991. Post-processing system for echo integrator data. Journal of the Acoustical Society of America, 90: 37–47.

Forbes, S.T., Nakken, O. 1972. Manual of methods for fisheries resource and appraisal. Part 2. The use of acoustic instruments for fish detection and abundance estimation. FAO Manual in Fisheries Science, 5. 138 pp.

Gerhardsen, T. 1946. Sildeleting ved hjelp av asdic og ekkolodd [Searching for herring using asdic and echo sounder]. Teknisk ukeblad, 93(51): 3–7. (In Norwegian).

Gjøsæter, H., Dommasnes, A., Røttingen, B. 1998. Acoustic Investigations of Size and Distribution of the Barents Sea Capelin Stock 1972–1997. Fisken og havet No 4 1998: 1–71.

Iversen, T. 1936. Sydøst Grønland og Jan Mayen. Fiskeriundersøkelser. [Southeast Greenland and Jan Mayen. Fishery investigations]. Fiskeridirektoratets Skrifter, Serie Havundersøkelser 5 (1): 1–171. (In Norwegian).

Johansen, K.E. 1989. Men der leikade fisk nedi havet: Fiskarsoge for Hordaland, 1920–1990 [Out in the Sea the fish played; The history of the fisheries in Hordaland 1920–1990] Bergen, J.W. Eide, 1989: 262 pp. (In Norwegian).

Knudsen, H.P. 1990. The Bergen Echo Integrator: an introduction. Journal du Conseil International pour l'Exploration de la Mer, 47: 167–174.

Korneliussen, R.J. 2002. Analysis and presentation of multifrequency echograms. Dr.scient thesis, Department of Physics, University of Bergen, Bergen, Norway.

Lea, E. 1947. ASDIC. Fiskets Gang 1947, No 1. (In Norwegian).

MacLennan, D.N. 1990. Acoustical measurement of fish abundance. Journal of the Acoustical Society of America, 87: 1–15.

Margetts, A.R. (Ed.). 1977. Hydro acoustics in Fisheries Research. Rapports et Procès-Verbaux des Réunions du Conseil International pour l'Exploration de la Mer, 170, 327 pp.

Midttun, L. 1966. Note on the measurement of target strength of fish at sea. ICES CM 1966: F: 9.

Midttun, L., Hoff, I. 1962. Measurement of the reflection of sound by fish. Fiskeridirektoratets Skrifter, Serie Havundersøkelser, 13(3): 1–18.

Midttun, L., Nakken, O. 1968. Counting of fish with an echo integrator. ICES Council Meeting 1968: B17.

Midttun, L., Nakken, O. 1971. On acoustic identification, sizing and abundance estimation for fish. Fiskeridirektoratets Skrifter, Serie Havundersøkelser, 16: 36–48.

Midttun, L., Nakken, O. 1977. Some results of abundance estimation studies with echo integrators. Rapports et Procès-Verbaux des Réunions du Conseil International pour l'Exploration de la Mer, 170, 253–258.

Midttun, L., Sætersdal, G. 1957. On the use of echo sounder observations for estimating fish abundance. Special publication. International Commission for the Northwest Atlantic Fisheries, 11: 260–266.

Midttun, L., Vestnes, G. 1977. Akustikk i fiskeriforskningen. [Acoustics in fisheries research]. Årsberetning for Havforskningsinstituttet [Annual report Institute of Marine Research] 1977. (In Norwegian).

Midttun, L., Sætersdal, G., Vestnes, G. 1957. Rapport om tokt med "G.O. Sars" til Nord-Norge og Barentshavet 25. februar–15. april 1957. [Report on cruise with "G.O. Sars" to Northern Norway and the Barents Sea, 25 February–15 April 1957]. Fiskets Gang no. 22, 1957. (In Norwegian).

Misund, O.A. 1993. Dynamics of moving masses: variability in packing density, shape and size among herring, sprat and saithe schools. ICES Journal of Marine Science, 50: 145–160.

Misund, O.A. 1997. Underwater acoustics in marine fisheries and fisheries research. Reviews in Fish Biology and Fisheries 7: 1–34.

Misund, O.A., Aglen, A., Johannessen, S.Ø., Skagen, D., Totland, B. 1993. Assessing the reliability of fish density estimates by monitoring the swimming behaviour of fish schools during acoustic surveys. ICES Marine Science Symposia, 196: 202–206.

Monstad, T. 2004. Blue whiting. In: Skjoldal, H.R. (Ed.): The Norwegian Sea Ecosystem. Tapir Academic Press, Trondheim. 2004: 263–288.

Nakken, O., Olsen, K. 1977. Target strength measurements of fish. Rapports et Procès-Verbaux des Réunions du Conseil International pour l'Exploration de la Mer, 170: 52-69.

Nøttestad, L., Fernö, A., Axelsen, B.E. 2002. Digging in the deep: killer whales' advanced hunting tactics. Polar Biology (2002): 939–941.

Olsen, K. 1971. Orientation measurements of cod in Lofoten obtained from underwater photographs and their relation to target strength. ICES CM 1971/B: 17.

Olsen, K. 1986. Sound attenuation within schools of herring. ICES CM 1986/B: 44.

Olsen, K. 1990. Fish behaviour and acoustic sampling. Rapports et Procès-Verbaux des Réunions du Conseil International pour l'Exploration de la Mer, 189: 147–158.

Olsen, K., Angell, J., Løvik, A. 1983. Quantitative estimations of the influence of fish behaviour on acoustically determined fish abundance. FAO Fisheries Report 300, 139–149.

Ona, E. 1982. Mapping the swim bladder's form and form stability for theoretical calculations of acoustic reflection from fish. Cand.real. thesis, University of Bergen, Bergen, Norway.

Ona, E. 1990. Physiological factors causing natural variations in acoustic target strength of fish. Journal of the Marine Biological Association of the United Kingdom, 70: 107–127.

Ona, E. 1994. Recent developments of Acoustic Instrumentation in Connection with Fish Capture and Abundance Estimation. In: Fernö, A., Olsen, S. (eds): Marine Fish Behaviour in Capture and Abundance Estimation. Fishing News Books, 1994: 221 pp.

Rollefsen, G. 1966. Norwegian fisheries research. Fiskeridirektoratets Skrifter, Serie Havundersøkelser, 14(1): 1–36.

Røttingen, I. 1976. On the relation between echo intensity and fish density. Fiskeridirektoratets Skrifter, Serie Havundersøkelser, 16: 301–314.

Runnstrøm, S. 1937. Sildeundersøkelser 1937 [Herring investigations 1937]. Årsberetning vedk. Norges Fiskerier 1937. (In Norwegian).

Runnstrøm, S. 1941. Quantitative investigations on herring spawning and its yearly fluctuations at the west coast of Norway. Fiskeridirektoratets Skrifter, Serie Havundersøkelser, 6(8): 1–71.

Sætersdal, G. 1955. Tokt med "G.O. Sars" til Barentshavet 22.7.–10.8.1955 [Cruise on "G.O. Sars" to the Barents Sea 22 July–10 August 1955]. Fiskets Gang, 38: 497–498. (In Norwegian).

Sætersdal, G., Hylen, A. 1959. Skreiundersøkelsene og skreifisket 1959 [The skrei investigations and the skrei fishery in 1959]. Fisken og Havet 1: 20 pp. (In Norwegian).

Scherbino, M., Truskanov, M.D. 1966. Determination of fish concentration by means of hydroacoustic apparatus. ICES CM 1966/F: 3.

Schwach, V. 2004. An Eye into the Sea. In: Rozwadowski, H. and van Keuren, D.K. (eds.): The Machine in Neptune's Garden: Historical Perspectives on Technology and the Marine Environment. Science History Publications/USA. Watson Publishing International.

Simmonds, E.J., Williams, N.J., Gerlotto, F., Aglen, A. 1992. Acoustic survey design and analysis procedure: A comprehensive review of current practice. ICES Cooperative Research Report 187.

Sogner, K. 1997. God på bunnen. SIMRAD-virksomheten 1947–1997 [The history of SIMRAD, 1947–1997]. (In Norwegian).

Sund, O. 1935. Echo sounding in fishery research. Nature, 135: 473–475.

Sund, O. 1939. Biological and oceanographic investigations. Cod stocks in 1938. Fish movements and replacements. Årsberetning vedk. Norges Fiskerier, 1938 (2): 87–102.
Sverdrup, H.U., Johnson, M.W., Fleming, R.H. 1942. The Oceans. Prentice-Hall, Inc. 1942: 1087 pp.
Toresen, R. 1991. Absorption of acoustic energy in dense herring schools studied by the attenuation in the bottom echo signal. Fisheries Research, 10: 317–327.
Vabø, R. 1999. Measurement and correction models of behaviourally induced biases in acoustic estimates of wintering herring. Dr. scient. thesis, Department of Biology, University of Bergen.
Zhao, X., Ona, E. 2003. Estimation and compensation models for the shadowing effect in dense fish aggregations. ICES Journal of Marine Science, 60: 155–163.

Fish behaviour, selectivity and fish-capture technology

Steinar Olsen

8.1 Introduction

The objective of fisheries management is to promote sustainable yields of harvestable natural resources while safeguarding the biodiversity of the aquatic environment. This requires in the first place fishing technology that facilitates efficient harvesting of the desired target species and sizes without destroying other organisms or damaging the capacity of the aquatic environment to sustain further life and growth. Secondly, rational and effective resource management must be based on adequate knowledge of the exploited stocks: their distribution and abundance, sex and age structure, etc. The research necessary to provide such knowledge invariably include catching fish in order to obtain biological samples and/or catch rates for fish density estimates. Thirdly, insight into and real understanding of the operations and capture processes of different types of fishing gear are indispensable for the formulation of effective, workable management regulations. Such knowledge can only be obtained by observations and measurements of fishing gear in actual operation, and by direct studies of relevant fish behaviour and reactions.

The three international fishing gear congresses organised by the Food and Agriculture Organization of the United Nations, FAO, were major events in the process of establishing research and development of fish-capture technology and related fish behaviour studies as a permanent field of fisheries science. The first of these, held in 1957 in Hamburg, was the eye-opener (Kristjonsson 1959), in that it made the authorities aware of the need for systematic studies and experiments, and of the great potential benefits of such work to the fishing industry. Research and development relevant to fish capture, including gear selectivity, were subsequently given higher priority by a number of fishing nations, as is evident from the reports submitted to the 'Second World Fishing Gear Congress', held 1963 in London (Finn 1964). One example was the establishment, initiated in Norway, of an informal international group with participants from six European nations, which in the 1960s attempted to organise an international forum for promoting cooperation in fish capture research across national borders (Dale and Møller 1964). This probably had a notable impact on the International Council for the Exploration of the Sea, ICES, to give fish behaviour and fish capture research higher priority than before, and to transform the previous 'Comparative Fishing Committee' into the current 'Fish Capture Committee'.

The next two outstanding international gatherings for promoting research and development in fish behaviour and fishing technology were the 1967 'FAO Conference on Fish Behaviour in Relation to Fishing Techniques and Tactics' in Bergen (Ben-Tuvia and Dickson 1968), and the 'Third Congress on Fishing Gear', organised by FAO in Reykjavik in 1970 (Kristjonsson 1971). Thereafter, twenty-two years elapsed before another major international event in this field took place, the 1992 ICES symposium in Bergen on 'Fish Behaviour in Relation to Fishing Operations' (Wardle and Hollingworth 1993). During the ensuing years, fish capture and behaviour research expanded greatly and became firmly and permanently established in Norway too.

8.2 The establishment of permanent fish capture research

Although Norway's first fishery research vessel, "Michael Sars", was equipped with a wide range of sampling gears in order to overcome their individual selective catching biases (Murray and Hjort 1912), prior to World War II, no directed, systematic studies of fishing gear and relevant fish behaviour were carried out in Norway.

Only incidental observations relevant to this multi-disciplinary field of research were made from time to time. However, the concept of size selection in fishing nets of different mesh sizes was always evident within the fishery research establishment. Thus, at the ICES Special Meeting in June 1934, captain Thor Iversen confirmed that trawl meshes do remain open so that small fish may escape, and he advocated the use of larger meshes in the Barents Sea trawl fisheries because of the large size of the fish there (Anon. 1934).

In 1953, Gunnar Rollefsen had a rare opportunity to compare the size distributions of gillnet- and longline-caught cod in Lofoten with those of catches taken by a presumably non-selective fishing gear, the purse seine. These unique data still makes his report (Rollefsen 1953) a classic publication on fishing gear selectivity. They were subsequently used by Sidney Holt (1963) to develop

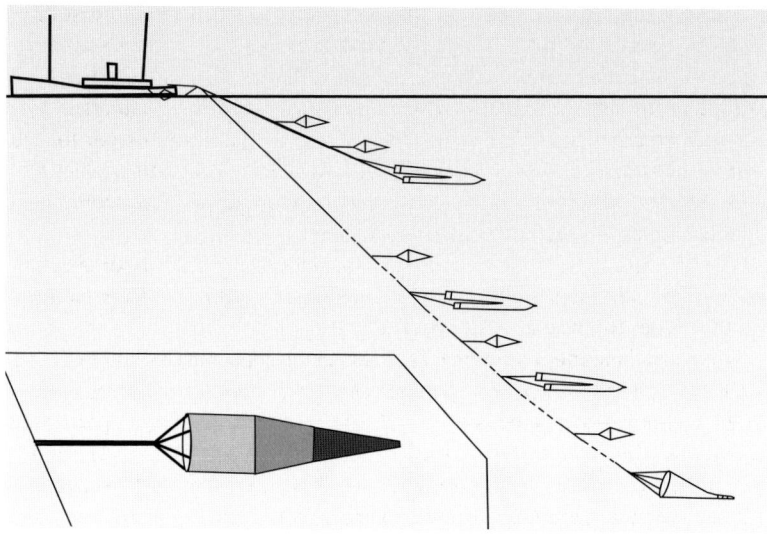

Figure 8.1
The sampling gears of "Michael Sars". (Redrawn from Murray and Hjort 1912)

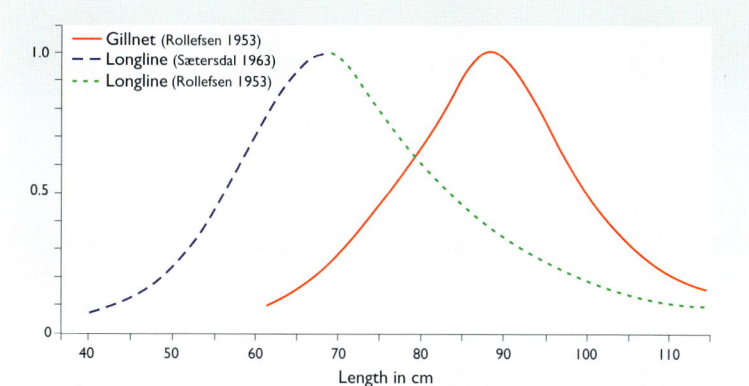

Figure 8.2
Selection curves for cod gillnets and longlines established from data of Rollefsen (1953) and of Sætersdal (1963).

mathematical models for gillnet mesh selection and new experimental methods. A dome-shaped gillnet size selection curve for cod was confirmed, and thereafter, using Holt's methods, also for herring (Olsen 1959) and halibut (Olsen and Tjemsland 1961, 1963). When Rollefsen's data are combined with those of Sætersdal (1963) for smaller cod, a dome-shaped long-line selection curve for the entire commercial size range for cod becomes apparent.

Of course, fishermen were always fully aware of the need to match the mesh size of their nets to that of the fish, and Rollefsen (*op. cit.*) observed that the gillnet fishermen in Lofoten tried to adjust the mesh size in their nets to correspond with the modal size of fish available. However, by basing these adjustments on the size of the fish they had taken in the previous year, they were usually out of phase with the annual progression in modal sizes caused by fluctuations in year-class strength.

After the Second World War, the need for effective management of fish resources became more and more apparent, and contemporary developments in the theory of fish population dynamics were gradually brought into use by fisheries biologists. It then became evident that quantitative knowledge was lacking with regard to the size-selection properties of fishing gears, in particular how the mesh size of a trawl affects the size of the fish caught. In the 1950s, therefore, several fisheries research institutions began to conduct extensive trawl mesh selection experiments. In the North-East Atlantic, this and related types of studies soon became coordinated through the ICES Comparative Fishing Committee. The Research and Statistics Committee of the International Commission for the Northwest Atlantic Fisheries, ICNAF, was also actively promoting selectivity work. The Institute of Marine Research, IMR, contributed to these coordinated international efforts right from their start.

The first milestone in the Norwegian investigations of trawl mesh selection was Gunnar Sætersdal's alternate haul experiments with RV "Thor Iversen" and RV "Peder Rønnestad" (Sætersdal 1955).

Sætersdal thereafter participated actively in the cooperative studies of ICES and ICNAF (Anon. 1961) to determine selection factors for different species of fish and how trawl selectivity is affected by various gear and operational parameters. This work was continued by Steinar Olsen and Arvid Hylen. However, these were studies of an *ad hoc* nature, conducted as and when required, without attempting to make an in-depth analysis of the interplay between fish

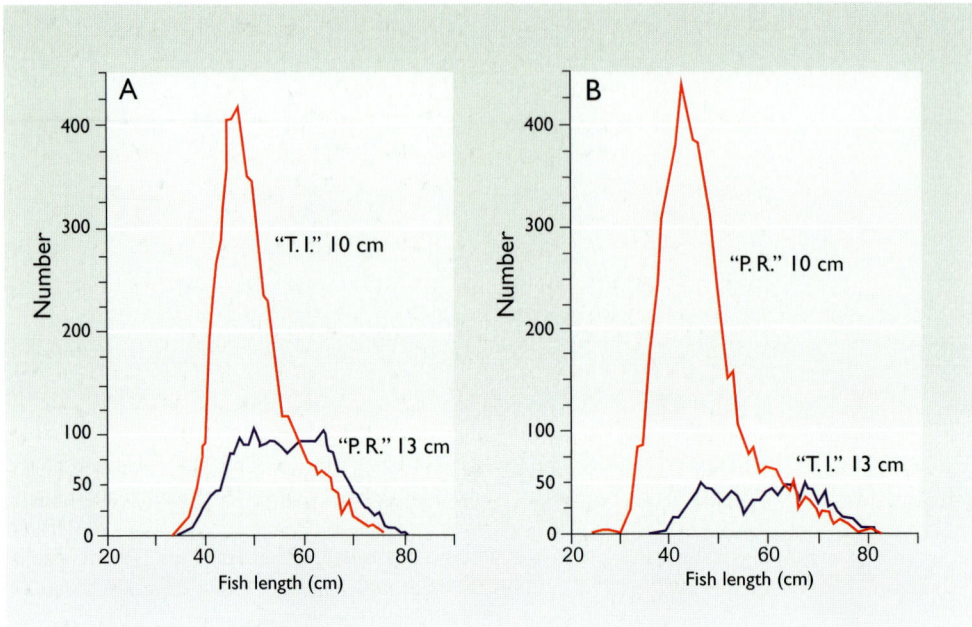

Figure 8.3
Results of alternate haul selectivity studies with RV "Thor Iversen" and RV "Peder Rønnestad". (Sætersdal 1955)

behaviour and fishing gear dynamics required to obtain greater insight into the relevant capture processes.

Sætersdal's early trawl selectivity studies included attempts to develop experimental methods for quantifying the relationship between codend mesh size and the sizes of the different kinds of fish caught, and to establish how this is affected by operational parameters such as catch size, towing time and speed. Subsequent investigations by Olsen and Hylen, while continuing these efforts, also responded to specific requests by the North East Atlantic Fisheries Commission for advice on the effects on trawl selectivity of new synthetic net materials, the use of various kinds of top-side chafers, discards (Hylen 1968), etc. These early studies of IMR thus made direct contributions to the international efforts of the time to generate quantitative knowledge of trawl mesh selectivity in the main commercial species of fish, and the effects on size selection by various fishing gear design features and operational parameters.

Subsequent needs to quantify the impacts on selectivity of new innovations in fishing technology and practices arose from time to time and had to be addressed, e.g. selectivity in pelagic trawls (Olsen 1975), and the effects on codend selectivity of round straps and new types of top-side chafers (Beltestad 1977). The greatly increased role of Danish seining in the coastal fisheries similarly necessitated new and better studies of size selection in this gear (Isaksen and Løkkeborg 1993).

At the beginning of the 1970s, the need for research and stock assessments increased greatly in order to meet the requirements of reliable management advice relevant to all the economically important fishery resources. This also called for more accurate and detailed insight into the capture processes in the main methods of commercial fishing. The desired fishing technology also had to be made less labour-intensive, more efficient and versatile, and capable of preventing, or at least greatly reducing, by-catch of undersized or unwanted kinds of fish. Accordingly, when in 1973 the new Institute of Fishery Technology Research (FTFI) was founded, it included a Fish Capture Division. This was located in Bergen in order to facilitate close cooperation with IMR and the Institute of Fishery and Marine Biology (IFMB) of the University of Bergen. During the 1980s, fish capture research capacity was also developed at the Universities of Tromsø and Trondheim. Particularly with regard to research on species and size selectivity in prawn and roundfish trawling, extensive cooperation was established with the Bergen group. Since 1974, continuous efforts have been made to enhance fish-capture technology in the Norwegian commercial fisheries, to reduce by-catch of undersized fish and non-target species, and to contribute to the reduction of bias and improved accuracy in research vessel survey and abundance estimates.

Throughout this period, studies of fish behaviour have been an integral part of all fish capture work. At the same time, important developments in underwater observation equipment and techniques have greatly enhanced the facilities for making direct *in situ* observations and measurements that had hitherto been inconceivable. A joint pool of instruments and equipment for fish behaviour studies was therefore established in Bergen, without which the progress made could not have been realised. Research has focused on virtually all methods of commercial fishing in Norway, with an emphasis on the fisheries for cod, haddock, prawn, herring and mackerel. In the following sections, we review work of direct relevance to cod and herring in some detail.

8.3 Enhancing fish capture technology

In the 1960s, a working group was established to develop with government funding a new, efficient single-boat pelagic trawl (Dale and Møller 1964). A number of basic research projects were instigated by this working group, but the funding authorities discontinued the extensive activities of the group before a workable trawl had been developed.

During the same period, public funds were also subsidising the initial development of an automatic baiting machine and its testing in the offshore longline fleet. When the Fish Capture Division of FTFI was established in 1974, its first task was to test a new longline baiting machine and compare its performance with that of the first one. This marked the beginning of an extensive and long-lasting engagement in research and development of longlining technology, always in close collaboration with the gear-manufacturing industry and the fishing fleet.

Direct tank and field observations of cod attacking baited hooks suggested ways of improving hook designs and clearly demonstrated the need for improving hooking probability by using a better hook than the traditional J-hook (Huse 1979). Better alternatives are now in general use in coastal longlining, and a completely new hook design has been developed in cooperation with the industry to fulfil the special requirements of autolining.

Detailed studies of how catching efficiency is also affected by the visibility of snoods and/or the mainline, twisting and effective length of snoods and by operational parameters, have provided a vast store of new knowledge and insight that, together with the better hooks and mechanised gear handling and baiting, has made modern longlining a high-tech industrial fishery (Bjordal and Løkkeborg 1996). Also noteworthy is the invention of a bird scaring device that

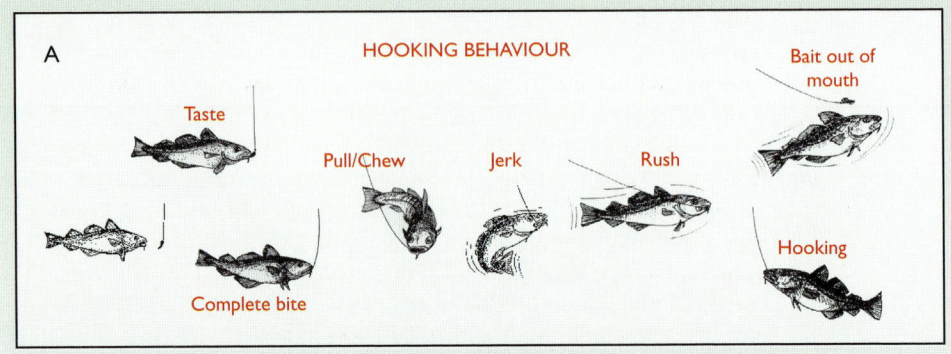

Figure 8.4a *Hooking behaviour in cod longline fishing. (From Bjordal and Løkkeborg 1996)*

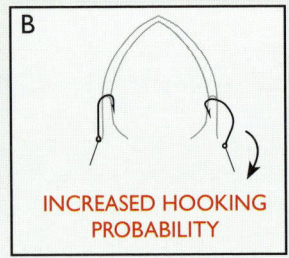

Figure 8.4b *Improved hook design. The wide gap hook is more likely to penetrate the tissue of the mouth cavity. (From Bjordal and Løkkeborg 1996)*

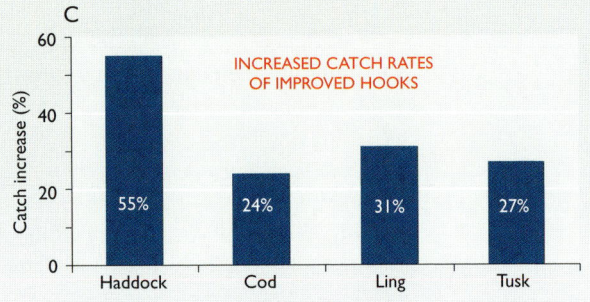

Figure 8.4c *Increased hooking probability and catch rates. (From Bjordal and Løkkeborg 1996)*

greatly reduces the incidental hooking and mortality of seabirds in long line fishing (Løkkeborg 1998).

In the coastal fisheries, seasonal changes in the fishing methods employed by individual vessels were, and still are, quite common, resulting in differences in the number of crew required on the vessels. Cod gillnetting, thus, was more labour-intensive than, for example, longlining. In order to abate this problem, increased mechanisation to reduce the manpower required for hauling gillnets was therefore attempted. This led to the introduction of smooth float and lead lines, and Angelsen *et al.* (1979) documented that both catching efficiency and size selectivity are affected by the hanging ratio and method of mounting the webbing of the net, as well as by the amount and distribution of buoyancy in the floatline. Hylen and Jakobsen (1979) found significant differences in catching efficiency for cod between gillnets made of different types of synthetic twine materials, but little effect on size selection.

Gillnet catching efficiency may also be improved by baiting the net, e.g. with a small net bag of longline bait attached in the middle of the net webbing (Engås *et al.* 2000). A gillnet simulation model developed by Dickson (1989), indicates that the main cause of this increase in catch rate is greater swimming activity in the vicinity of the bait bag. However, this inexpensive method to increase the catch rate in round fish gillnetting has not yet come into general use in commercial fishing.

The first Research and Development research (R&D) project of the Fish Capture Division explored the possibility of baited trap fishing for groundfish in Norwegian coastal waters. Initially traps similar to those used for black cod on the west coast of the USA were tested. Good catches of tusk were sometimes obtained, but not of cod or other gadoids, and the traps were found to be too large and cumbersome to be operated by the Norwegian coastal fleet (Valdermarsen 1976). In 1987, experiments started to test a collapsible light-weight fish trap. This proved easy to operate, once again producing good catches of tusk, but not of cod (Bjordal and Furevik 1988). Subsequent underwater TV studies revealed that cod easily escape from this simple type of trap. A new two-chamber trap design gave much improved cod catch rates (Furevik and Hågensen 1997), and is now regarded as an alternative commercial gear for the cod fishery.

Another pioneering project of the new Fisheries Technology Research Institute continued the work already started by IMR on developing suitable trawl gear for the new blue whiting fishery that had started in 1974. In order to obtain a sufficiently large trawl mouth area to obtain catches large enough for profitable commercial fishing with the towing powers available, larger mesh sizes in the front part of the trawl were tested. In 1975 a trawl with 2 m meshes in the front part was tested, in 1978 the biggest meshes were 10 m, and during the following two years, new designs with up to 30 m elongated hexagonal meshes were tried. It was shown that to maintain relatively low towing resistance in the large pelagic trawls required for catching blue whiting, very large meshes could be used in the forward part without significant escapes of blue whiting (Isaksen *et al.* 1979). Subsequent gear developments were made by the industry itself, and at present maximum mesh sizes up to 124 m are used in the largest trawls with mouth openings of 150 m in height and 200 m wide.

In 1976, experiments with hexagonal shaped meshes (H-net) in purse seine webbing started, and this type of net proved to have advantages over standard net with diamondshaped meshes both with regard to required hanging ratio, sinking speed and hydrodynamic resistance. Accordingly, H-net has been much used in coastal purse seines for saithe and in North Sea herring and capelin seines (Beltestad 1980). For the very largest herring seines, however, the H-net is said to have insufficient tearing strength.

Other research activities in purse seine technology and relevant target fish behaviour, have included trials with reduced hanging ratio, construction and mounting of the float and lead lines, total lead weight and distribution in relation to sinking speed, and the use of large-meshed webbing in the end of the seine without excessive escapes of caught fish (Beltestad, Dickson and Misund 1988). In order to reduce manpower requirements in coastal purse seining, a better ring needle and net hauling and mechanical stacking system were developed. Subsequently, development of a mechanised net stacking system for large offshore purse seiners was started. When initial encouraging results had been obtained, the industry itself took over further development and completion of the project.

8.4 New concepts in solving the by-catch problems

Given that trawl mesh size regulations alone were not effectively preventing catches and the destruction of huge numbers of undersized fish, and that in many trawl fisheries large by-catches of unwanted non-target species were also being caught and discarded, for the past thirty years Norwegian fish capture research efforts have therefore been aimed specifically at these problems.

When the Norwegian trawl fishery for deep-water prawn, *Pandalus borealis*, was first developed off the southern coast during the early 20th century (Hjort and Ruud 1935), it was quickly realised that as well as prawns, the catches included large by-catches of small fish. On the southern shrimp grounds, however, these largely consisted of non-commercial fish species, hence, the only management regulations were imposed: a minimum codend mesh size of 36 mm and a minimum depth limit of 100 m for prawn trawling were aimed at reducing catches of small prawns. However, when prawn fishing expanded to North-Norway and greatly increased in volume, it was observed with much concern, that the by-catch discarded was largely made up of small cod and haddock with little or no chance of survival. Local regulations banning prawn trawling in many North-Norwegian coastal localities were therefore imposed.

Nevertheless, the problem persisted, and the additional mortality of small cod generated by the coastal prawn fishery was tentatively estimated to significantly reduce subsequent fishery yields in the cod fishery. Tests were therefore started by IMR (Rasmussen 1973) on a shrimp separation trawl system developed in the USA (Jurkovitch 1970). This was based on an original French design for shallow-water shrimp (Kurk and Faure 1965), utilising the different behaviour of shrimp and small flatfish.

Experiments in the IMR circular tank confirmed that *Pandalus borealis*, like shallow water shrimp, are passively filtered in a trawl net (Johannessen 1976), and a new separator trawl was developed and thoroughly tested. This design incorporated an obliquely mounted net in front of the codend with a mesh size large enough to let the shrimp pass through into the codend. An opening in the trawl net just above the separating panel allows the fish to escape (Karlsen and Mathai 1978). The performance of this trawl system was found to be satisfactory in general, but it was quite sensitive to installation errors (including wilful tampering), catch size and composition, as well as methods of operation. Evidently, a more reliable prawn separation technology was required.

Partly inspired by the invention of Paul Solberg, a fisherman from Møre, the installation of a metal grate in his prawn trawl to reduce jellyfish by-catch, but also by that of the 'Turtle Extruder Device', TED, developed for the US shrimp fishery in the Mexican Gulf (Watson and Seidel 1980); the separator net panel was replaced by a rigid metal grid. Here the prawn pass between the metal bars of the grid into the codend, while fish are guided towards an escape opening above. This design proved to be a break-through, releasing practically all fish above 18–20 cm and significantly reducing the catch of smaller ones as well. The performance of the system is very little affected by prawn catch size and/or operational parameters such as towing speed and duration (Isaksen *et al.* 1992). In 1990, therefore, it was made compulsory for prawn fishing in Norwegian coastal waters, and from 1991 also for the offshore prawn fishing in the Norwegian EEZ and the Svalbard Fish Protection Zone, and as from 1993 in the Russian EEZ.

While the effects of these prawn fishing management regulations on recruitment to the exploitable stock of cod have not been evaluated in detail, the introduction of the grid selection system in the Norwegian prawn fisheries is estimated to have reduced the level of discards of under-sized/unwanted fish (mainly cod, haddock and redfish) by around 90 per cent (Anon. 2000), while in the coastal region, where by-catch regulations in the prawn fisheries have been

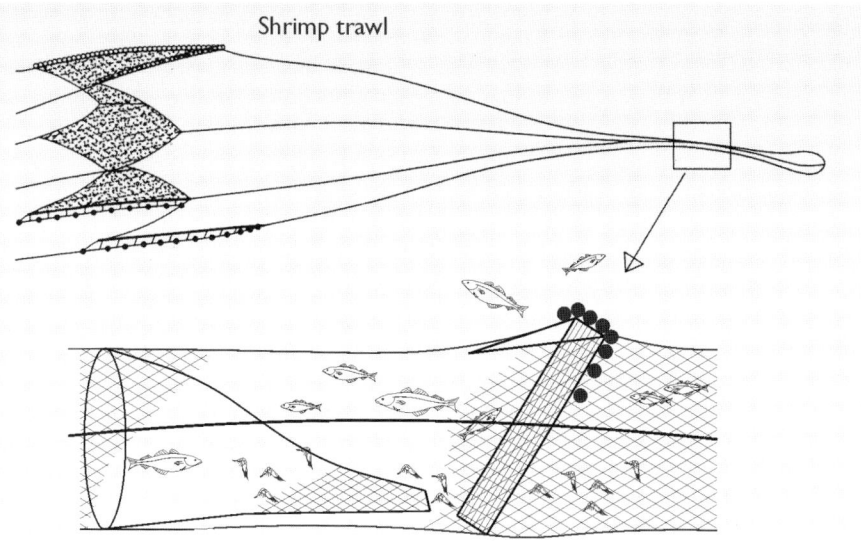

Figure 8.5
Illustration of a by-catch excluder grid in a shrimp trawl. (From Isaksen and Valdemarsen 1994)

enforced for the longest time, the abundance of cod subsequently increased. In addition to the impacts of strong cod year-classes during the same period, this increase may also have been an effect of reduced cod mortality prior to their recruitment to the commercial stock. In the offshore shrimp fisheries, it is conceivable that today's more responsible prawn fishing technology may have had an even greater impact on pre-recruitment mortality of cod and other roundfish.

The conspicuous success in reducing fish by-catch in the prawn trawl fisheries by a rigid metal selection grid, germinated the idea of also improving fish size selection by means of a grid. Initial experiments at the University of Tromsø were encouraging (Larsen 1990), and in cooperation with IMR, a grid system for size selection of roundfish was eventually developed (Larsen and Isaksen 1993). This incorporates two hinged metal grids with a given bar spacing, installed in the extension piece in front of the codend, and an escape opening for small fish cut in the trawl panel above the grids. The selection performance of this system was found to be stable and very little affected by towing speed, catch size or composition. It was therefore made compulsory in all roundfish bottom-trawl fishing by Norwegian vessels in the national EEZ north of 62°N. However, when severe and dangerous problems of handling the grid were experienced in rough

Figure 8.6
The single-grid trawl system for size selection of roundfish.

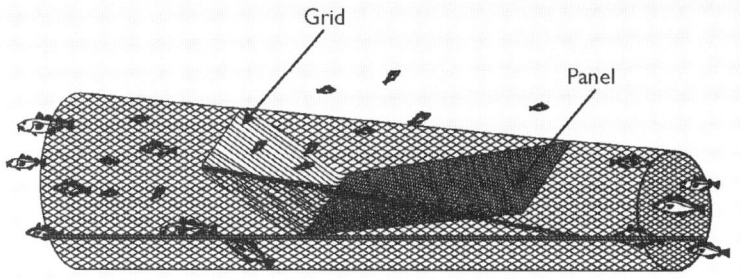

weather, compulsory utilisation was made weather-dependent. A lighter single-grid system was developed and tested in cooperation between IMR and PINRO, Murmansk (Isaksen *et al.* 1998). Russia is enforcing the use of this selection system on her trawlers, and it has also been authorised for use by Norwegian trawlers operating north of 62°N.

As a follow-up of the encouraging Scottish experiments in the North Sea, codends with square-shaped meshes were tested for cod and haddock in the Barents Sea (Isaksen and Valdemarsen 1988). These tests confirmed that selectivity is markedly improved in the square mesh codend, the selection range is reduced, and selectivity is little affected by towing speed and catch size. However, because of increased meshing of redfish in the square meshes and other operational problems with square mesh codends, this innovation was at the time found less useful in the Barents Sea. Work therefore proceeded with other ideas to provide more open meshes in a standard codend. This was achieved by shortening the lastridge ropes on each side, or by transferring much of the longitudinal strain to the lower panel of the codend by making it a few meshes shorter than the top one (Isaksen and Valdemarsen 1990). However, when fishermen learned about this success, knowing also that cod and haddock usually escape upwards, some of them simply installed the codend upside down with the shortened panel in the top part!

In contrast to trawls, no operational problems were experienced when using square mesh codends in Danish seine (Isaksen *et al.* 1997), and they were easier to handle than selection grids. Selectivity studies confirmed that both devices released small cod and haddock as well as, or slightly better than, bottom trawls.

Engås *et al.* (2000) showed that by modifying the water flow in the trawl codend escapes of small red snapper were greatly enhanced. This discovery may lead to better selection in future prawn trawls also of smaller sizes than those released today, e.g. of redfish and 0-group cod.

A study has started on utilising biological sounds from fish to develop more species- and size-selective fishing methods, in the first place cod, haddock and tusk. Recorded sequences of fish sound will be replayed to the same species, and any attractive and repulsive behavioural responses observed will be utilised to improve the selectivity of existing passive fishing methods (Soldal *et al.* 2000).

In quota-regulated demersal fisheries, which with few exceptions are mixed fisheries of several kinds of fish, problems arise when the quota of one species has been taken. However, research and development of species-selective finfish

Figure 8.7
System to modify the water flow in the trawl codend to enhance selection of small fish. (From Engås et al. 2000)

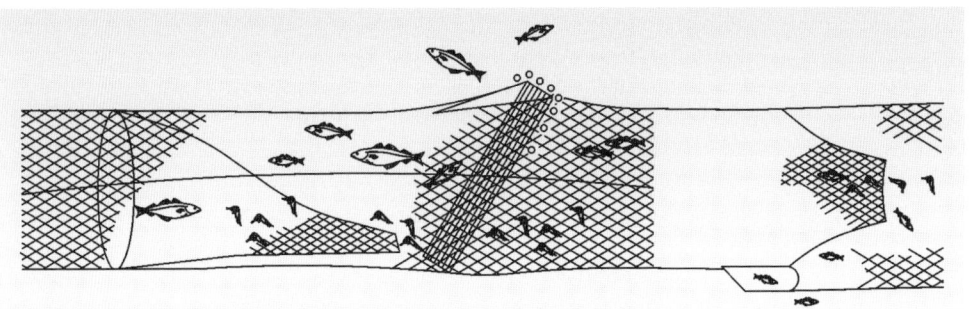

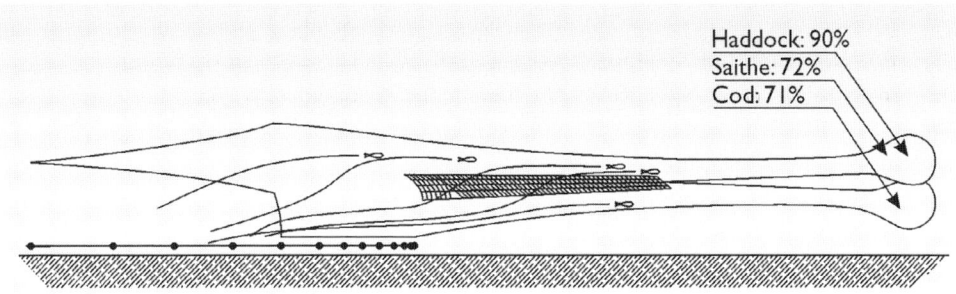

Figure 8.8
*Sorting trawl with horizontal square mesh panel.
(From Isaksen and Waldemarsen 1988)*

fishing methods have until recently been given rather low priority and consequently made little progress. To be able to target one particular species of fish, or, conversely, avoid catching some undesired fish, tends to be quite difficult because the differences in behaviour and reactions to fishing gear between the various kinds of fish are often minor and not easily recognised. Nevertheless, the small and subtle differences that do exist can in some cases be utilised. In bottom trawl fishing, cod and haddock differ in their escape reaction (Main and Sangster 1983), and the differences have been exploited in the development of a trawl that is horizontally divided by a square mesh panel and has two separate codends. In this trawl as much as 90 per cent of the haddock catch was taken in the upper codend, while 60 per cent of the cod ended up in the lower one (Engås *et al.* 1998).

In special cases, size selection alone may result in distinct species selection. Thus, by-catch of haddock in Danish seining for plaice at the Finnmark coast was eliminated completely by using a square mesh codend of 170 mm mesh size, which also released all cod of less than 70 cm (Isaksen 1993).

The solution to severe by-catch problems may even necessitate conversion to alternate fishing methods. Two-chamber cod traps were thus found to keep cod and king crab separate in the traps, which allowed the crab by-catch to be easily released. Compared with gillnet fishing at the same period at the same sites, equivalent catch rates of cod were maintained (Furevik and Hågensen 1997).

Visibility, colour and texture of the twine may affect gillnet catching efficiency differently for different kinds of fish (Engås and Løkkeborg 1994), thus enhancing species selection. In longline fishing, both the type and size of baits may affect both species and size selection, as to some extent does the type of hook used (Løkkeborg 1994).

8.5 Unaccounted mortality of escaping fish

Mesh-size regulations are enforced in many fisheries to prevent fishery-caused mortality of fish below a certain size or age, the assumption being that most of the undersized fish selected out by the meshes are alive and will survive and grow to reach commercial size. However, fish that come in contact with fishing gear, but subsequently escape, often suffer injuries that may impair their survival prospects. Wounds caused by gillnets and jiggers in cod have frequently been observed in coastal fisheries. Sundnes (1961, 1991) found that in the Lofoten cod fishery up to 49 per cent of the cod in individual purse seine landings were suffering from gear injuries, mainly caused by jiggers and gillnets, and that some of the lesions were very severe and probably lethal. Another cause of unaccounted

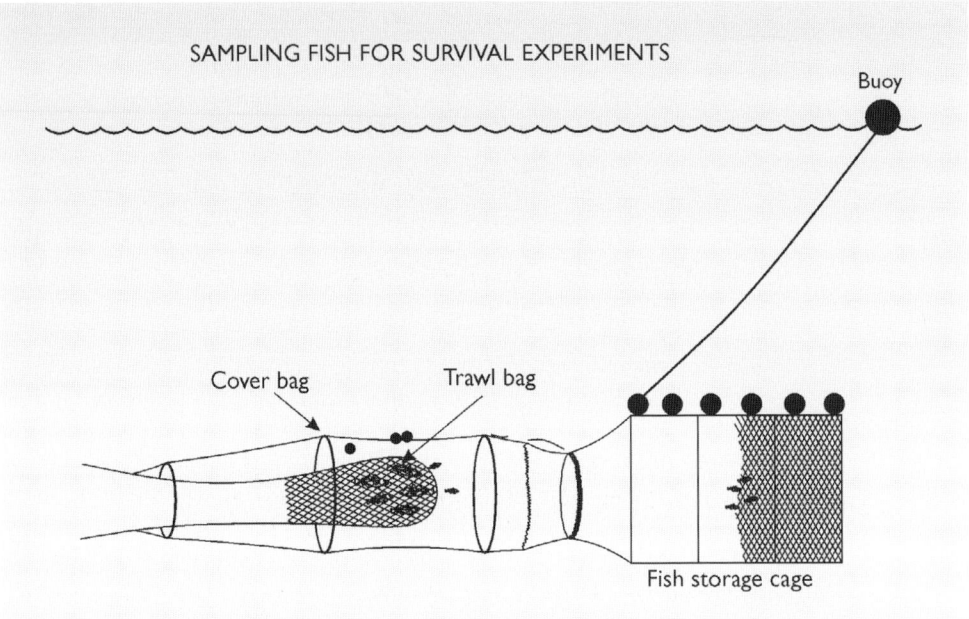

Figure 8.9
Experimental cage system to study survival of fish escaping through sorting grid. (From Isaksen et al. 1998)

fishing mortality in cod gillnet fishing is that fish caught in bottom-set gillnets may be destroyed by hagfish, amphipods and isopods (Angelsen and Engås 1982). They are therefore discarded and not recorded as being caught.

In longlining, unaccounted fishing mortality may be caused by undersized and unwanted fish being ripped off the hook at the rail and released in a wounded condition. Another cause is the loss of fish that become detached from the line near the surface and float to the surface belly up, unable to descend to the bottom (Bjordal and Chruikshank 1989; Soldal and Huse 1997). Improved hooks and mainlines with swivels (Bjordal 1989) have reduced the incidence of fish becoming detached near the surface and the problem may also be abated by adjusting the bait size, which is the most important size-selection factor in longlining (Løkkeborg and Bjordal 1992).

In the late 1980s, concern increased about the fate of fish that escape through trawl meshes. In Norway extensive experiments were therefore carried out by collecting fish that had escaped through trawl meshes in detachable large cages that surrounded the codend during towing (Soldal et al. 1993).

Little or no cod mortality was observed during the two weeks after capture, while a few percent of the haddock died during the observation period. However, Løkkeborg and Soldal (1995) later found that vulnerability to predation is greater in small cod that escape from a trawl. On the other hand, practically one hundred per cent survival was observed in young gadoids released from a shrimp trawl by a rigid deflecting grid (Soldal and Engås 1997).

The roundfish trawl industry, however, claimed that the IMR survival studies had not been conducted under realistic commercial fishing conditions, and that observations made by the trawler fleet suggested increased mortality of escaping young fish after the compulsory introduction of the grid selection system. New experiments were therefore carried out in August 2000. The results gave no evidence in support of the belief that grid selection in cod trawl fishing causes higher mortality in escaping undersized codfish than does trawling without the use of a grid-selection system (Soldal *et al.* 2001).

Clupeoids are very sensitive to contacts with fishing gear. Olsen (1981) observed large numbers of dead herring on the bottom in the vicinity of cod gillnets, and concluded that these had been caught by their teeth in the gillnet twine and subsequently succumbed ('the death bite').

Between 1985 and 1987 the total unaccounted fishing mortality of Norwegian spring-spawning herring was estimated at about 50 000 tonnes (Beltestad and Misund 1989). This high mortality was assumed to occur in connection with net bursts or when live herring were being stored in net pens. In the winter herring fisheries, net bursts occur most frequently during the day, when the purse seine is shot on large, dense schools migrating to the spawning grounds and the catch may exceed 300 tonnes. For some years only night fishing was therefore allowed during the winter herring fisheries. Studies of the survival of herring after simulated net bursts have indicated that the reason for the high mortality associated with net bursts is probably severe scale loss followed by lethal osmo-regulation difficulties (Misund and Beltestad 1995).

The smaller coastal purse seiners that take most of their herring catch in the autumn transfer the catches live into net pens for storage for up to several weeks before delivery. Significant mortality occurs here too, and the dead herring are usually discarded. Experiments clearly showed that this mortality is mainly caused by two factors: the speed of towing the net pens from the catch locality to the inshore mooring site, and the size of the net pens. Low towing speed (0.5 knots) and large net pens (> 1000 m^3) may therefore greatly reduce unaccounted fishing mortality in this fishery (Misund and Beltestad *op. cit.*).

8.6 Sample fishing and abundance estimation

The fishing operations carried out during research vessel surveys are basically intended to serve two objectives. One is to provide biological samples and establish the species and size composition of the fish present in the location, e.g. in acoustic surveys. The other is to obtain catch-rate data for estimating fish densities.

Much effort has been put into improving the sampling gears themselves and their methods of operation in order to provide more representative catches with low variance. Conversely, by identifying and quantifying variability and biases in catching performance, these factors can be taken into account and adjusted for in abundance estimates. More recently, focus has also been put on developing more optimal survey and sample fishing designs, e.g. by choice of towing time, frequency and distribution of tows with regard to area and time (Godø 1994).

Fish avoidance of vessels and gear has always been a fact of life in commercial fishing, and it has been studied for a long time as well in research vessel surveying (Godø *op. cit.*). Particularly in acoustic estimations of fish abundance, avoidance behaviour is a very serious source of bias. This is because the resulting errors, for example caused by sudden changes in tilt angle, may be large, and are difficult to detect properly (Olsen 1971; Olsen *et al.* 1983; Aglen 1994).

Both horizontal and vertical avoidance movements may seriously affect the catching efficiency of survey sampling gear, especially with regard to pelagic species. The magnitude of escape responses has been shown to vary over short

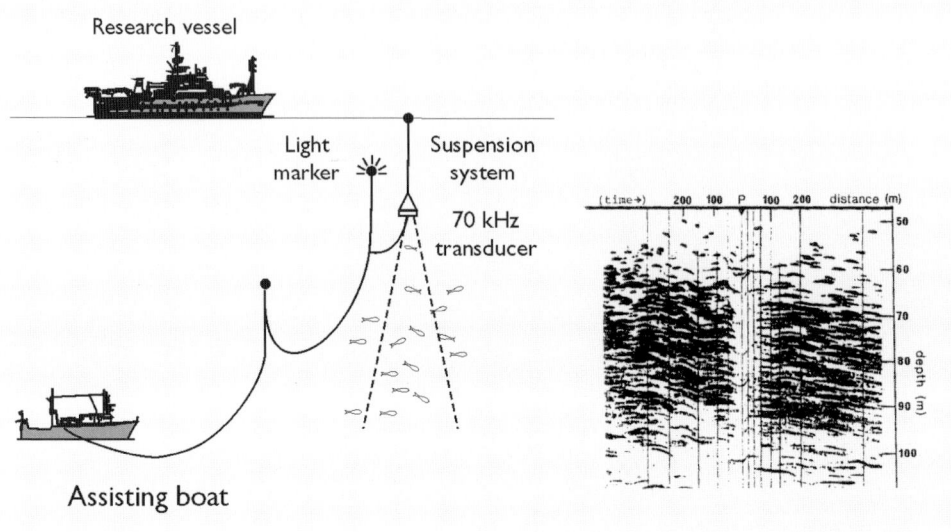

Figure 8.10 Studies of fish avoidance. (From Olsen et al. 1983)

periods of time and relatively small geographical areas. The problem is further confounded by differences in the level and characteristics of noise between individual research vessels, which may cause their inherent biases in catching efficiency to differ (Engås 1994).

The catch performance of a sampling trawl is also greatly affected by the trawl geometry, which should therefore remain stable during the tow as well as between tows. Door and wing spread, and vertical opening of the trawl in the Norwegian bottom trawl surveys were found to vary systematically with depth. To reduce the variability in trawl performance at different depths, a constraining rope mounted 150 m ahead of the doors was introduced. Compared with standard procedures, this method substantially reduced the variability of door spread with depth (Engås *op. cit.*).

Continuous instrument monitoring of the trawl and door bottom contact is essential to assess the effective distance towed. Recently developed instruments also facilitate consistent bottom contact by keeping towing speed through water, warp tension and warp elevation angle constant (Engås, *op. cit.*).

Sweep angle and sweep length significantly affect the catch performance and size selectivity of a sampling trawl. At great sweep angles, the sandclouds, which provide a strong herding stimulus, are propagated along a line outside the sweeps and wing-ends (Main and Sangster 1983). This enhances the ability of fish to escape by crossing over the sweeps. Engås and Ona (1990) measured distances of up to 7.5 m between the wingtips and the sandclouds in the Norwegian sampling trawl. While catch rates of larger cod and haddock were generally found to increase when sweep lengths were increased from 20 m to 120 m, smaller fish (< 30 cm) were clearly underrepresented in the catches (Engås, *op. cit.*).

Hylen *et al.* (1986) observed that the standard trawl surveys gave annual abundance estimates for pre-recruit age groups of cod that increased to age 3. Experiments with small-meshed net bags attached below the sampling trawl confirmed significant size- and species-dependent escapes of fish beneath the

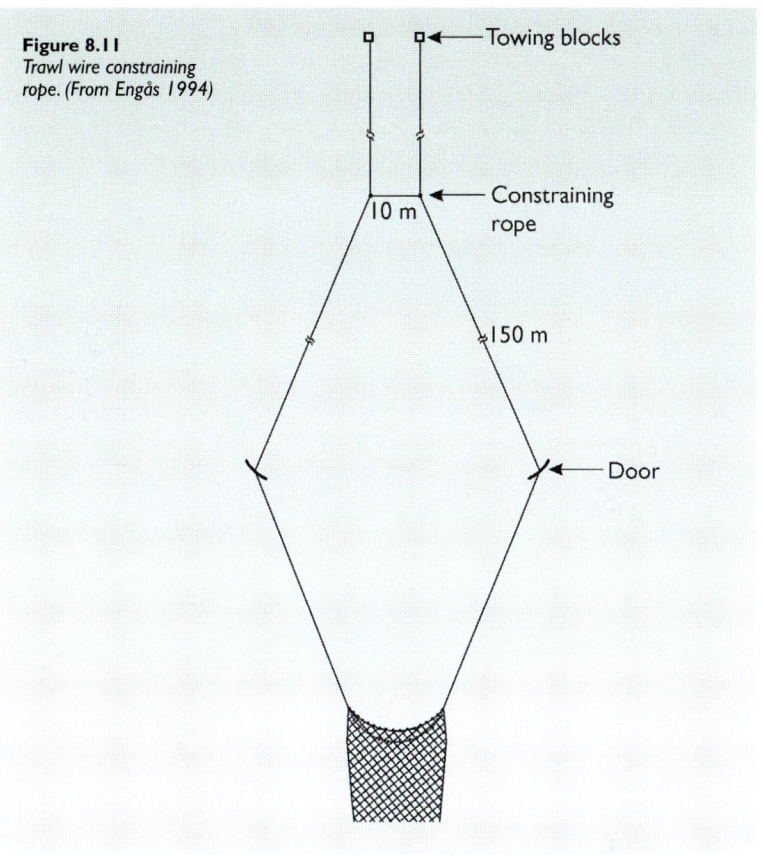

Figure 8.11
Trawl wire constraining rope. (From Engås 1994)

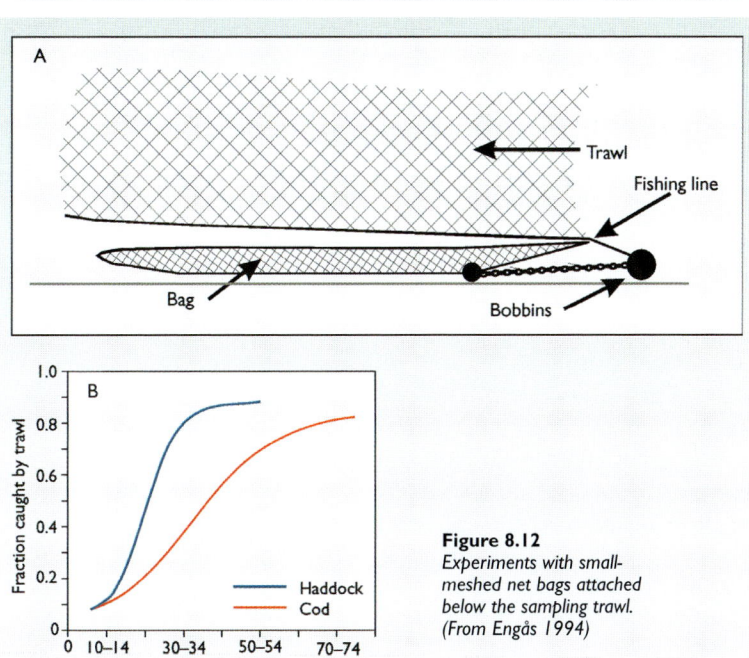

Figure 8.12
Experiments with small-meshed net bags attached below the sampling trawl. (From Engås 1994)

fishing line. Much of the resulting sampling bias was reduced by the introduction of heavy rock-hopper ground gear (Engås *op. cit.*), and the experiments have also provided data for adjusting the magnitude of the resulting bias (Godø 1994).

Norwegian and Canadian video and trawl efficiency experiments indicated that the density of ground fish in the area swept by a trawl may be positively related to the catchability of cod and other species (Godø *et al.* 1999). In contrast, Angelsen and Olsen (1987) demonstrated a negative relationship between fish density, fishing effort (e.g. local gear density), and catching efficiency in the Lofoten gillnet and longline cod fisheries.

Changes in the spatial distribution of cod and haddock stocks (Korsbrekke *et al.* 1993) have proved that standardised survey designs may lead to inefficient distribution of effort. This strongly underlines the need for better methods of allocating survey effort (Godø *op. cit.*). For example, monitoring the stock distribution acoustically enables the accuracy of survey estimates to be improved

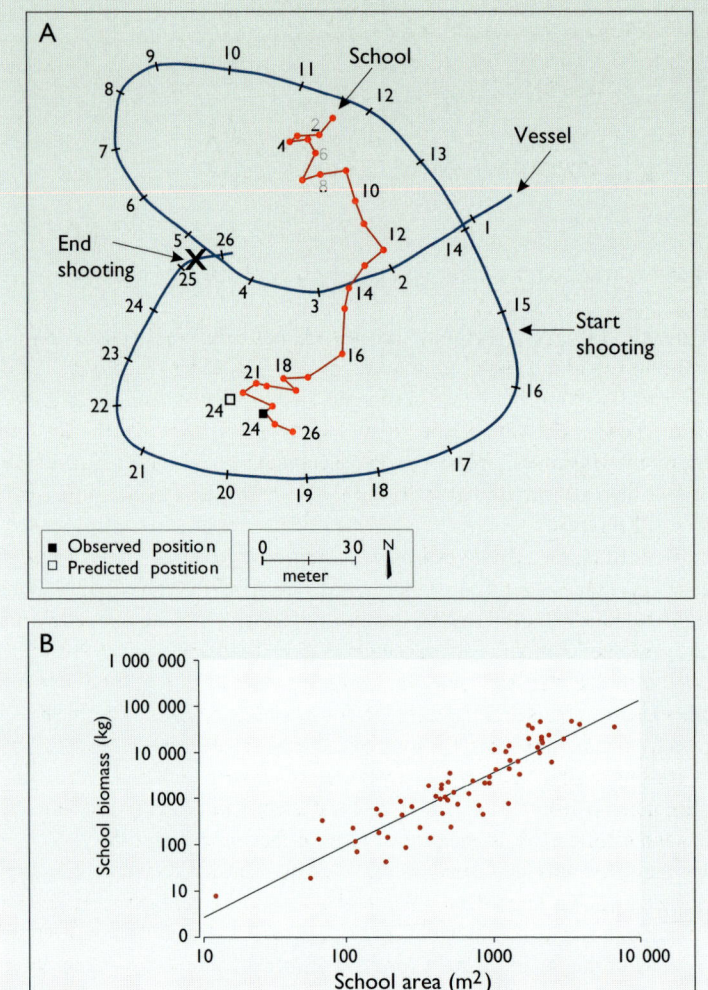

Figure 8.13
Sonar measurements of herring schools have provided data for better mapping and abundance estimates of schooling pelagic fish. (From Misund 1994)

A) School movement observed by sonar during a purse seine operation.

B) The relationship between school area measured by sonar and school biomass of the catch.

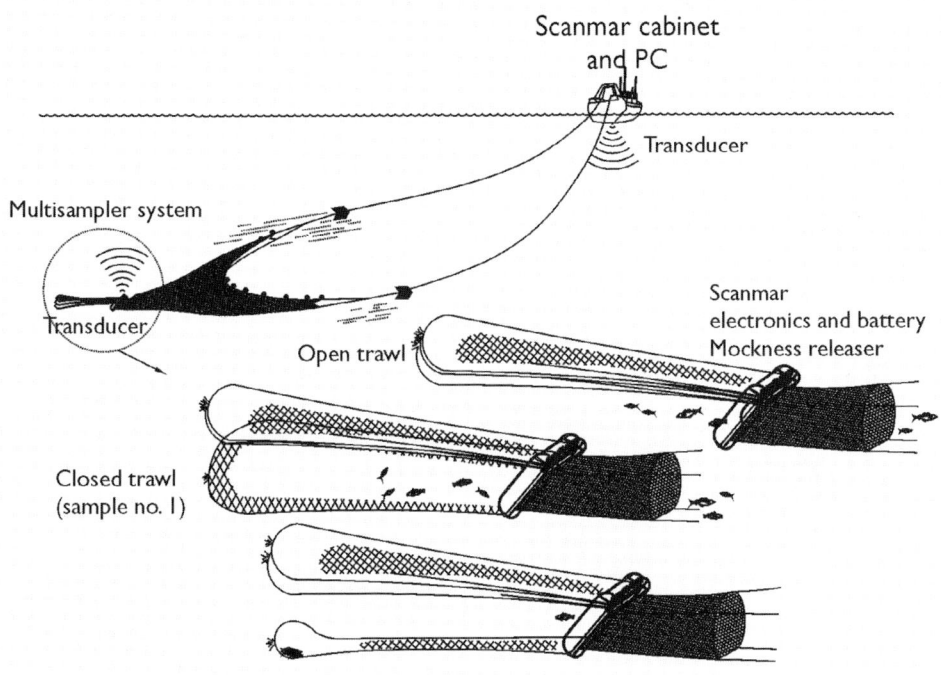

Figure 8.14
Multi-codend trawl system. (From Engås et al. 1987)

by selecting trawl stations during the survey according to a predetermined sampling scheme (Ona *et al.* 1991).

Available research vessel time may also be better utilised by reducing tow duration and working more trawl stations. Survey accuracy might thereby be improved without increasing costs. Estimates of population parameters would not be significantly affected by reducing tow duration. On the contrary, long tows may negatively bias abundance estimates. Godø (*op. cit.*) therefore advocated flexible tow durations by using additional information from catch sensors and acoustics.

Until recently, abundance estimates of herring have mainly been based on acoustic surveys that employ standard echo integration techniques. Misund (1991, 1993) studied swimming behaviour of herring schools in relation to capture gears and acoustic abundance estimation. He demonstrated that school dimensions measured using sonar were linearly related to biomass as determined by subsequent purse seine captures of the schools measured. The method is affected by the swimming behaviour of the schools (Misund *et al.* 1993) and has accordingly been refined by combining sonar data with echo integration measurements (Misund *et al.* 1995), and is currently becoming routinely used to map and estimate the abundance of near-surface schooling fish.

A system for remotely opening and closing multiple codends on pelagic sampling trawls has been developed to allow up to three separate samples to be collected during a single haul. This has provided data for estimating size selection as a function of distance towed and also, from pelagic trawl hauls, for species and size distribution by depth (Engås *et al.* 1997).

REFERENCES

Aglen, A. 1994. Sources of Error in Acoustic Estimation of Fish Abundance. In: Marine Fish Behaviour in Capture and Abundance Estimation (eds. A. Fernö & S. Olsen) Fishing News Books Ltd., Oxford.

Angelsen, K., Haugen, K., Floen, S. 1979. The catching efficiency of cod gillnets with different hanging ratio (E) and different floatline buoyancy. ICES CM 1979/B: 19.

Angelsen, K.K., Engås, A. 1982. Fish destroyed by amphipodes, a catch reduction factor in gillnet fishing. ICES FTFB Working Group, Aberdeen, May 1982.

Angelsen, K.K., Olsen, S. 1987. Impact of Fish Density and Effort Level on Catching Efficiency of Fishing Gear. Fisheries Research, 5 (1987) 271–278.

Anon. 1934. Size limits for fish and regulations of the meshes of fishing nets. ICES Special meeting, June 1932. Rapports et Procès-Verbaux des Réunions du Conseil International pour l'Exploration de la Mer, XC.

Anon. 1961. Report of the Mesh Selection Working Group's Meeting in Copenhagen, December 1959 and 1960. ICES CM 1961, Comparative Fishing Committee.

Anon. 2000. Report of the FTFB Topic Group on Unaccounted Mortality in Fisheries. ICES FTFB Working Group, Haarlem, The Netherlands, 10–11 April 2000.

Beltestad, A.K. 1977. Selectivity experiments with topside chafers and round straps. ICES CM 1977/B: 38.

Beltestad, A.K., 1980. Recent experiments with net of hexagonal meshes in purse seines. ICES CM 1980/B: 25.

Beltestad, A.K., Misund, O.A. 1989. Is unaccounted fishing mortality a problem in purse seining? ICES FTFB Working Group meeting, Dublin, 24–26 April 1989.

Beltestad, A.K., Dickson, W., Misund, O.A. 1989. Optimization of purse seines by large meshed sections and low lead weight. Theoretical considerations, sinking speed measurements and fishing trials. Proceeding World Symposium on Fishing Gear and Fishing Vessel Design, St. John's, Newfoundland, November 1988, pp. 527–530.

Ben-Tuvia, Dickson, W. (eds) (1968). Proceeding of the Conference on Fish Behaviour in Relation to Fishing Techniques and Tactics, Bergen, Norway, 19–27 October 1967. FAO Fisheries Report, 62.

Bjordal, Å. 1989. Recent developments in longline fishing – catching performance and conservation aspects. Proceeding World Symposium on Fishing Gear and Fishing Vessel Design, St. John's, Newfoundland, 25 November 1988. Pp. 19–24.

Bjordal, Å., Furevik, D.M. 1988. Full-scale fishing trials for tusk (*Brosme brosme*) and cod (*Gadus morhua*) with a collapsible fish trap. ICES CM, 1988/B: 33.

Bjordal, Å., Chruikshank, O. 1989. Fish escapement from longlines and methods to study escapement and survival. ICES WG meeting, Dublin, 24–26 April 1989.

Bjordal, Å., Løkkeborg, S. 1996. Longlining. Fishing News Books Ltd. Oxford.

Dale, P., Møller, S. 1964. The Development of a Midwater Trawl. In: Modern Fishing Gear of the World, Vol. 2 (ed. H. Kristjonsson, Fishing News (Books) Ltd., London 1964.

Dickson, W. 1989. A simulation model for cod gillnet effectiveness. Proceeding World Symposium on Fishing Gear and Fishing Vessel Design, St. John's, Newfoundland, November 1988, pp. 60–65.

Engås, A. 1994. Catching Efficiency of Demersal Sampling Trawls. In: Marine Fish Behaviour in Capture and Abundance Estimation (eds. A. Fernö & S. Olsen). Fishing News Books Ltd., Oxford.

Engås, A., Ona, E. 1990. Day and night fish distribution pattern in the mouth area of the Norwegian bottom-sampling trawl. Rapports et Procès-Verbaux des Réunions du Conseil International pour l'Exploration de la Mer, 187, 123–127.

Engås, A., Løkkeborg, S. 1994. Abundance Estimation using Bottom Gillnet and Longline. In: Marine Fish Behaviour in Capture and Abundance Estimation (eds. A. Fernö & S. Olsen), Fishing News Books Ltd., Oxford.

Engås, A., Skeide, R., West, C.W. 1997. The MultiSampler: a system for remotely opening and closing multiple codends on a sampling trawl. Fisheries Research 29 295–298.

Engås, A., Jørgensen, T., West, C.W. 1998. A species-selective trawl for demersal gadoid fisheries. ICES Journal of Marine Science, 55: 835–845.

Engås, A, Jørgensen, T., Angelsen, K.K. 2000. Effects on catch rates of baiting gillnets. Fisheries Research 45, 265–270.

Engås, A., Foster, D., Hataway, B.D., Watson, J.W., Workman, I. 2000. The behavioural Response of Juvenile Red Snapper (*Lutjanus campechanus*) to Shrimp Trawls that Utilize Water Flow Modifications to Induce Escapement. Marine Technology Society Journal. Vol. 33, No 2. 43.

Finn, D.B. 1964. Modern Fishing Gear of the World, Vol 2. Fishing News Books Ltd. Oxford.

Furevik, D., Hågensen, S.P. 1997. Forsøk med torsketeiner som alternativ til garn i Varangerfjorden i perioden april–juni og oktober–desember 1996. Prosjekt nr. 06071. Institute of Marine Research, Bergen, June 1997. (In Norwegian).

Godø, O.R. 1994. Groundfish Abundance Estimates from Bottom Trawl Surveys. In: Marine Fish Behaviour in Capture and Abundance Estimation (eds. A. Fernö & S. Olsen), Fishing News Books Ltd., Oxford.

Godø, O.R., Walsh, S.J., Engås, A. 1999. Investigating density-dependent catchability in bottom-trawl surveys. ICES Journal of Marine Science, 56: 292–298.

Hjort, J., Ruud, J.T. 1938. Rekefisket som naturhistorie og samfundssak. Fiskeridirektoratets Skrifter, Serie Havundersøkelser Vol. V, No 4. (In Norwegian).

Holt, S.J. 1963. A method for determining gear selectivity and its application. Special Publication, International Commission, NW Atlantic Fisheries. 163(5): 106–115.

Huse, I. 1979. Influence of hook design and gear materials in longlining for cod (*Gadus morhua* L.) and haddock (*Melanogrammus aeglefinus* L.) investigated by behaviour studies and fishing trials. Cand.real. thesis. University of Bergen, Bergen, Norway (In Norwegian).

Hylen, A. 1968. Discarding of fish in the North-East Atlantic. ICES Cooperative Research Report, Series B.

Hylen, A., Jacobsen, T. 1974. A fishing experiment with multifilament, monofilament and monotwine gillnets in Lofoten during the spawning season of Arcto-Norwegian cod. Fiskeridirektoratets Skrifter, Serie Havundersøkelser; 16: 531–550.

Hylen, A., Nakken, O., Sunnanå, K. 1986. The use of acoustic and bottom trawl surveys in the assessment of North-East Arctic cod and haddock stocks. Pp. 473–498. In: M. Alton (Ed.). A Workshop on Comparative Biology, Assessment and Management of Gadoids from the North Pacific and Atlantic Oceans, Seattle Washington, June 1985.

Isaksen, B. 1993. Fangst og mellomlagring av rødspette. Rapport fra Senter for marine ressurser nr. 22–1993. Havforskningsinstituttet. (In Norwegian).

Isaksen, B., Valdemarsen, J.W. 1988. Selectivity experiments with square mesh codend in bottom trawl, 1985–1987. Workshop on the Selectivity of Square Mesh in Trawls, World Symposium on Fishing Gear and Fishing Vessel Design, St. John's, Newfoundland, 25 November 1988.

Isaksen, B., Valdemarsen, J.W. 1990. Codend with short lastridge ropes to improve size selectivity in fish trawls. ICES CM 1990/B: 46.

Isaksen, B., Valdemarsen, J.W., Larsen, R.B., Karlsen, L. 1992. Reduction of fish by-catch in shrimp trawl using a rigid separator grid in the aft belly. Fisheries Research, 13: 335–352.

Isaksen, B., Løkkeborg, S. 1993. Escape of cod (*Gadus morhua*) and haddock (*Melanogrammus aeglefinus*) from Danish seine codends during fishing and surface hauling operations. ICES Marine Science Symposia, 196: 86–91.

Isaksen, B., Gamst, K., Misund, R. 1997. Sammenligning av bruks- og seleksjonsegenskaper hos sorteringsrister og kvadratmasker for snurrevad. Intern rapport, Havforskningsinstituttet, mars 1997. (In Norwegian).

Isaksen, B., Gamst, K., Kvalsvik, K., Axelsen, B. 1998. Comparisons of selection and user features of SORT-X and single sorting grids as used in a two-panel cod trawl. IMR Report, January 1998.

Johannessen, A. 1976. Tank observations of prawns and small cod in relation to a moving trawl. ICES CM 1976: B: 29.

Jurkovitch, J.E. 1970. Shrimp-fish separator trawls with a method of modifying a Gulf of Mexico shrimp trawl for use in waters of the states of Oregon and Washington. Conference on Canadian shrimp fishery, St. John's, Newfoundland, 27–29 October 1970.

Karlsen, L., Mathai, J. 1978. Experiments with separating panels in coastal shrimp trawls in Norway in March and October/November 1977. FTFI-report.

Korsbrekke, K., Mehl, S., Nakken, O., Nedreaas, K. 1993. Investigation on demersal fish in the Barents Sea, winter 1993. Rapport fra Senter for marine ressurser, 14–93, Institute of Marine Research, Bergen, Norway.

Kristjonsson, H. (ed.) 1959. Modern Fishing Gear of the World, Vol. 2. Fishing News Books Ltd. Oxford.

Kristjonsson, H. (ed.) 1971. Modern Fishing Gear of the World, Vol. 3. Fishing News Books Ltd. Oxford.

Kurc, G., Faure, L. 1965. Un nouveau modèle de chalut sélectif pour la pêche de crevettes. ICES CM 1965. Paper No 65.

Larsen, R.B. 1990. Testing a new sorting technology for bottom trawls to avoid catch of juvenile fish. Norwegian College of Fisheries Science, University of Tromsø, Report 10 pp. (In Norwegian).

Larsen, R.B., Isaksen, B. 1993. Size selectivity of rigid sorting grids in bottom trawls for Atlantic cod (*Gadus morhua*) and haddock (*Melanogrammus aeglefinus*). ICES Marine Science Symposia 196, 178–182.

Løkkeborg, S. 1994. Fish Behaviour and Longlining. In: Marine Fish Behaviour in Capture and Abundance Estimation (eds. A. Fernö & S. Olsen), Fishing News Books Ltd., Oxford.

Løkkeborg, S. 1998. Seabird by-catch and bait loss in longlining using different setting methods. ICES Journal of Marine Science, 55: 145–149.

Løkkeborg, S., Bjordal, Å. 1992. Species and size selectivity in longline fishing: a review. Fisheries Research, 13: 311–322.

Løkkeborg, S., Soldal, A.V. 1995. Vulnerability to predation of small cod *(Gadus morhua)* that escape from a trawl. ICES CM 1995/B:15 Ref. G. 7 pp.

Main, J., Sangster, G.I. 1983. Fish reactions to trawl gear - A study comparing light and heavy ground gear. Scottish Fisheries Research Report, 27.

Misund, O.A. 1991. Swimming behaviour of schools related to fish capture and acoustic abundance estimation. Dr.philos.thesis, University of Bergen, Dept. of Fisheries and Marine Biology, Bergen, Norway.

Misund, O.A. 1993. Abundance estimation of fish schools based on relationship between school area and school biomass. Aquatic Living Resources, 1993 (6): 235–241.

Misund, O.A., Aglen, A., Johannessen, S.Ø., Skagen, D., Totland, B. 1993. Assessing the reliability of fish density estimates by monitoring the swimming behaviour of fish schools during acoustic surveys. ICES Marine Science Symposia, 196, 202-6.

Misund, O.A., Beltestad, A.K. 1995. Survival of herring after simulated net bursts and conventional storage in net pens. Fisheries Research, 22 (1955), 293–297.

Misund, O.A., Aglen, A., Hamre, J., Ona, E., Røttingen, I., Valdemarsen, J.W. 1995. Mapping of schooling fish near the surface by sonar, echo integration and surface trawling. ICES Fisheries and Plankton Acoustics Symposium, Aberdeen, 12–16 June 1995, 14 p. 2 tables.

Murray, J., Hjort, J. 1912. Depths of the Ocean. Macmillan and Co., London.

Olsen. K. 1971. Influence of vessel noise on the behaviour of herring. In: Modern Fishing Gear of the World, Vol. 3 (ed. H. Kristjonsson) pp. 291–294. Fishing News Books, Ltd., Oxford.

Olsen, K. 1984. By-catch of herring in cod gillnets. Contribution to ICES Working Group on Fisheries Technology and Fish Behaviour.

Olsen, K., Angel, J., Peterson, F., Livid, A. 1983. Observed fish reaction to a surveying vessel with special reference to herring, cod, capelin and polar cod. In: Symposium on Fisheries acoustics. Selected Papers of the ICES/FAO Symposium on Fisheries Acoustics. Bergen, Norway, 21–24 June 1982 (eds. O. Nakken & S.C. Venema). FAO Fisheries Report 300, 131–8.

Olsen, S. 1959. Mesh selection in herring gill nets. Journal of Fisheries Research Board of Canada, 16: 339–349.

Olsen, S., Tjemsland, J. 1961. The Selectivity of Halibut Gill Nets. ICES CM 1961/18.

Olsen, S., Tjemsland, J. 1963. A method of finding an empirical total selection curve for gill nets, describing all methods of attachments. Fiskeridirektoratets Skrifter, Serie Havundersøkelser, 13 (6): 88–94.

Olsen, S. 1975. Selectivity of pelagic trawls. ICES WG meeting, Oostende, 21–25 April 1975.

Ona, E., Pennington, M., Vølstad, J.H. 1991. Using acoustics to improve the precision of bottom trawl indices of abundance. ICES CM 1991/D: 13.

Rasmussen, B. 1973. Fishing experiments with selective shrimp trawls in Norway, 1970–1973. FAO Fisheries Report, 139, 56.

Rollefsen, G. 1953. The selectivity of different fishing gear used in Lofoten. Journal du Conseil International pour l'Exploration de la Mer 19: 191–194.

Soldal, A.V., Engås, A., Isaksen, B. 1993. Survival of gadoids that escape from a demersal trawl. ICES Marine Science Symposia, 196: 122–127.

Soldal, A.V., Engås, A. 1997. Survival of young gadoids excluded from a shrimp trawl by a rigid deflection grid. ICES Journal of Marine Science, 54: 117–124.

Soldal, A.V., Huse, I. 1997. Selection and mortality in pelagic longline fisheries for haddock. ICES CM 1997/FF: 14.

Soldal, A.V., Midling, K.Ø., Fosseidengen, J.E., Svellingen, I., Øvredal, J.T. 2000. Fish sound, a future tool for selectivity. In: Proceedings Fifth European Conference on Underwater Acoustics, Lyon, 10–13 July 2000. (Ed. M.E. Zakharia).

Soldal, A.V., Isaksen, B., Gamst, K. 2001. Dødelighetforsøk med torsketrål sommeren 2000. Statusrapport til "Ordningen for fiskeforsøk og veiledningstjeneste", Institute of Marine Research, Bergen, January 2001. (In Norwegian).

Sundnes, G. 1961. Forsøksfiske med not i Lofoten i 1961. Undersøkelser over redskapsskadd skrei. Directorate of Fisheries, Bergen. (Unpuplished, mimeographed, in Norwegian).

Sundnes, G. 1991. Redskapsskader på den norsk-russiske torskestammen. Det Kgl. Norske Videnskabers Selskabs Forhandlinger 1991. (In Norwegian).

Sætersdal, G. 1955. Maskeviddeforsøk med småtrål, mai 1954. Fiskeridirektoratets småskrifter, nr. 4, 1955. (In Norwegian).

Sætersdal, G. 1963. Selectivity of longlines. Special Publication, International Commission, NW Atlantic Fisheries. 1963(5): 189–192.

Valdemarsen, J.W. 1976. Norwegian experiments with deep-sea traps. ICES CM 1976/B: 7.

Wardle, C., Hollingworth, C.E. (eds.) 1993. Fish Behaviour in Relation to Fishing Operations. ICES Marine Science Symposia, Vol. 196.

Watson, J.W., Seidel, W.R. 1980. Development of a selective trawl for the southeastern United States shrimp fishery. ICES CM 1976/B: 28.